ROUTLEDGE LIBRARY EDITIONS:
20TH CENTURY SCIENCE

Volume 10

SCIENTISTS AS WRITERS

SCIENTISTS AS WRITERS

Edited by
JAMES HARRISON

Routledge
Taylor & Francis Group

LONDON AND NEW YORK

First published in 1965

This edition first published in 2014
by Routledge
2 Park Square, Milton Park, Abingdon, Oxon, OX14 4RN

and by Routledge
711 Third Avenue, New York, NY 10017

Routledge is an imprint of the Taylor & Francis Group, an informa business

First issued in paperback 2016

© 1965 James Harrison

All rights reserved. No part of this book may be reprinted or reproduced or utilised in any form or by any electronic, mechanical, or other means, now known or hereafter invented, including photocopying and recording, or in any information storage or retrieval system, without permission in writing from the publishers.

Trademark notice: Product or corporate names may be trademarks or registered trademarks, and are used only for identification and explanation without intent to infringe.

British Library Cataloguing in Publication Data
A catalogue record for this book is available from the British Library

ISBN13: 978-1-138-01357-5 (hbk)
ISBN13: 978-1-138-98147-8 (pbk)

Publisher's Note
The publisher has gone to great lengths to ensure the quality of this book but points out that some imperfections from the original may be apparent.

Disclaimer
The publisher has made every effort to trace copyright holders and would welcome correspondence from those they have been unable to trace.

Scientists as Writers

Edited by JAMES HARRISON

METHUEN AND CO LTD

11 NEW FETTER LANE, LONDON EC4

First published 1965
© *James Harrison* 1965
Printed in Great Britain by
W. & J. Mackay & Co Ltd, Chatham

Acknowledgements

I wish to thank the following for permission to use material of which they hold the copyright:

GEORGE ALLEN & UNWIN, LTD., for extracts from *Evolution: the Modern Synthesis* by Julian Huxley, and *The ABC of Relativity* by Bertrand Russell.

G. BELL & SONS, LTD., for an extract from *An Approach to Modern Physics* by E. N. da C. Andrade.

ERNEST BENN, LTD., for an extract from *Scientific Autobiography and Other Papers*, by Max Planck (tr. Frank Gaynor), published by Williams & Norgate.

CAMBRIDGE UNIVERSITY PRESS, for extracts from *The Nature of the Physical World* by Arthur Eddington, *The Stars in their Courses* by James Jeans, *Man on his Nature* by Charles Sherrington, *The Two Cultures and the Scientific Revolution* by C. P. Snow, and *Science and the Modern World* by A. N. Whitehead.

CASSELL & CO., LTD., for an extract from *What Man May Be* by George Russell Harrison.

THE CLARENDON PRESS, OXFORD, for extracts from Aristotle's *Metaphysica*, tr. W. D. Ross, and *Doubt and Certainty in Science* by J. Z. Young.

THE CRESSET PRESS, LTD., for extracts from *Man on Earth* by Jacquetta Hawkes.

GERALD DUCKWORTH & CO., LTD., for an extract from *The Living Brain* by W. Grey Walter.

FABER & FABER, LTD., for an extract from *The Rape of the Earth* by G. V. Jacks and R. O. Whyte.

VICTOR GOLLANCZ, LTD., for extracts by J. Bronowski and Bertrand Russell from *What is Science*, edited by James R. Newman.

WILLIAM HEINEMANN, LTD., for extracts from *The Common Sense of Science* by J. Bronowski, and *Science Past and Present* by F. Sherwood Taylor.

THE HOGARTH PRESS, LTD., for an extract from *New Introductory Lectures on Psycho-Analysis* by Sigmund Freud (tr. W. J. H. Sprott).

HUTCHINSON & CO. (PUBLISHERS), LTD., for extracts from *Physics and Microphysics* by Louis de Broglie (tr. Martin Davidson), *The Physicist's Conception of Nature* by Werner Heisenberg (tr. Arnold J. Pomerans), and *Error and Deception in Science* by Jean Rostand (tr. Arnold J. Pomerans).

THE LOEB CLASSICAL LIBRARY for an extract from Diogenes Laertius' *Lives of Eminent Philosophers*, tr. R. D. Hicks.

THE OBSERVER, and the Author, for an extract from an article by John Davy.

PENGUIN BOOKS, LTD., for an extract from *Human Disease* by A. E. Clark-Kennedy.

LAURENCE POLLINGER, LTD., for an extract from *Silent Spring* by Rachel Carson (published in U.K. by Hamish Hamilton).

THE UNIVERSITY OF NORTH CAROLINA PRESS, for extracts from *Science and Christian Belief* by C. A. Coulson (published in U.K. by O.U.P. and Fontana).

YALE UNIVERSITY PRESS, for an extract from *On Understanding Science* by James B. Conant (published in U.K. by O.U.P.).

Contents

CONTENTS

Introduction

This book consists of a selection of prose passages by scientists about science, which I have arranged around eleven main themes, and to which I have added a minimum of linking comment and annotation. For those who wish to use them, there are suggestions at the end of the book for discussion, for written work, and for further reading.

The book is intended for use in all those General Courses, under whatever name they go, in Sixth Forms, Technical Colleges and Training Colleges – courses where it is hoped the students will sharpen their wits, improve their command of English, and at the same time broaden the over-specialized field of study covered by their other subjects.

The selection does not pretend to add up to a history and philosophy of science; nevertheless, anyone who studies it carefully cannot help but learn something about the state of science yesterday and today, and, more important, will have been led to consider how science works and what it can and cannot do. There is no need, in our twin-cultured society, to stress why the education of those who are not going to be scientists should include more of this kind of thinking about science. All too often, however, I believe the need to be almost as great among those who are studying to be scientists.

Most of the main themes which the extracts have been chosen to illustrate will, in any case, introduce discussions and opportunities for written work on topics ranging far beyond the confines as they are usually conceived of science.

I take this opportunity of thanking all those of my colleagues who have helped and encouraged me in any way during the preparation of this book.

Shenstone College,
Bromsgrove.   J. H.

CHAPTER ONE

The Nature of the Universe

1. Greek Astronomy

The earliest Greeks seem to have thought the earth to be flat or cylindrical, but, from the fifth century on they knew that the earth was a sphere, and nearly all of them supposed this sphere to be in the centre of the universe and motionless. The stars were supposed to be at the surface of a vast sphere concentric with the earth. The stars rise and set once a day, so this sphere was supposed to rotate on its axis once in 24 hours. Outside this there was nothing, or a sort of unformed chaos. The chief problem of antiquity was to work out how the bodies between the earth and stars moved. These were the moon and sun, and the planets, Mercury, Venus, Mars, Jupiter and Saturn. These rose and set, but not in exact time with the stars. The moon rises about an hour later every day, the sun about four minutes (reckoned by the stars). The planets behave very oddly, for although they generally very slowly fall behind the stars, they sometimes stay still relatively to them and even gain on them for a short time. The Greeks decided that the only *fitting* path for a heavenly body was the 'perfect' figure of a circle; and Aristotle, indeed, laid down that just as on earth bodies, if undisturbed, naturally fell down (i.e. towards the centre), so heavenly bodies *naturally* moved uniformly in circles round the centre.

The problem the Greeks set themselves was to discover a combination of circular motions which would give rise to the curious paths which the planets in fact took. The first important solution was that of concentric spheres. Imagine a 'nest' of spheres one inside the other, turning on axles like wheels. Each sphere has its axle set in the surface of the sphere next larger than itself. The axles incline in various directions and the spheres rotate at very different speeds. The planet, borne on the innermost sphere, will combine the motions of them all, and will describe at intervals a sort of figure-of-eight path which is not unlike its apparent path through the heavens. The correspondence of this system with the observed facts was not at all exact; and another difficulty was that the planets change in brightness and so seem sometimes to be nearer to the earth and sometimes further from it, which could not be the case if they moved on a sphere concentric with it.

The idea of concentric invisible spheres, adopted though not invented by Aristotle, persisted until the seventeenth century;[1] but at the same time another and better system, the Ptolemaic, was used by astronomers from the second to the seventeenth century. The earth, as before, is supposed to be motionless at the centre and the planets to move round it. Each planet moves round a circle, the epicycle, the centre of which moves in another circle[2] which surrounds the earth, though its centre is not at the earth's centre. The centre of the epicycle does not, however, move uniformly, but so as to lie always on the uniformly rotating radius of a third circle.[3] This complicated system, if the size of the circles, the position of their centres and the rate of rotation were properly chosen, could give a very fair account of the motion of the planets. This was important in all ages till about 1700, because men took much account of astrology, which foretold the events of their lives by the relative positions of the stars and planets. Although by modern standards these systems were far from accurate, they were a great advance on no system at all, and, moreover, were wonderful feats of mathematical reasoning.

The Greeks did not confine themselves to systems in which the earth stands still at the centre of the universe. Aristarchus of

---

[1] *These were the spheres famous for their music. Many people during the Middle Ages lost sight of the fact that a planet needed a whole nest to itself, and thought of there being just one sphere for each planet, including the sun and moon, with an eighth for the stars and sometimes a ninth outermost one. (Ed.)*

[2] *The deferent. (Ed.)*

[3] *i.e. The centre of the epicycle does not move at a uniform* speed *round the circumference of the deferent; in other words it moves so as to lie not on the uniformly rotating radius of the deferent, but on the uniformly rotating radius or spoke of a third circle, though free, so to speak, to slide up and down along its length. The centre (called the equant) of this third circle and the earth's centre lie equidistant on either side of the centre of the deferent. (Ed.)*

Samos, about 270 B.C., proposed a system identical with the Copernican; it attracted few, if any, followers, however, and there was talk of a charge of impiety being brought against him.

F. SHERWOOD TAYLOR, *Science Past and Present*, 1945

To say that the only 'fitting' path for a heavenly body is the perfect figure of a circle, or that stones have a 'desire' to reach the centre of the earth, or that nature 'abhors' a vacuum – all these strike us nowadays as too quaint, too anthropomorphic, to qualify as scientific statements. But in their time they were serious attempts to express certain observed facts of nature in a generalized form. And when closer observations showed the inadequacy of the first of them in its simplest form, the further attempt to plot planetary movements as combinations of several circles at least imposed some kind of coherence or order on what might otherwise have seemed wholly random scrawlings across the sky.

Ptolemy lived in the second century A.D., Aristarchus in the third B.C. Not until Copernicus (1473–1543) did the heleocentric hypothesis find another champion. And even then, Copernicus merely put it forward as another possible way of accounting for appearances, which was why he could number bishops among his patrons. It was Galileo's tactless insistence, not that this might be so but that this was so, which made it impossible for the Pope to ignore his challenge to the authority of church and scripture.

However, it was left to Newton to establish finally not only that this was so but how (and why?) this was so.

11. The Achievement of Newton

Men have known for several thousand years that the sun and the planets move in regular ways against a background of stars which seem to be still. These regularities can be used to look forward as well as back: the Babylonians were able to use them to forecast eclipses of the sun. The sun, the moon and the planets can be pictured as being carried round the earth on these regular paths in great shells or spheres. Or the paths, which seen from the earth are curiously looped, can be thought of as the rolling of wheels upon wheels; it was in this way that Ptolemy and other Greeks in Alexandria patterned them on the night sky eighteen hundred years ago. Ptolemy's picture does not claim to explain the movement of the planets, if indeed we could make him understand this meaning of the word 'explain' which has become natural for us. It gives an order to their movements by describing them, and so tells us where we may expect to see them next.

Two things happened in the sixteenth century to make astronomy ill at ease with this description; and they are both of interest, because they remind us that science is compounded of fact and logic. The Danish astronomer Tycho Brahe took better and more regular observations of the positions of the planets, and they showed that Ptolemy's paths, charming though they looked as mathematical curves, were really only rather crude guides to where the planets rolled. And even earlier, Copernicus showed that these paths were much simpler if they were looked at not from the earth but from the sun. Early in the seventeenth century, these two findings were combined by Kepler, who had worked for Brahe. Kepler used the measurements of Brahe and the speculations of Copernicus to frame general descriptions of the orbits of the planets: for example, he showed that, seen from the sun as focus, a planet sweeps out equal areas of its ellipse in each equal interval of time.

It was these empirical generalisations of Kepler which Newton and his contemporaries worked from when they began to look for a deeper order below the movements of the planets. They had also a new weapon of theory. For while Kepler had been at work in the north, Galileo in Italy had at last overthrown the physical conceptions in the works of Aristotle, which had long been attacked in Paris. By the time the Royal Society was founded, the complicated Greek ideas of motion with their conflict of earth and air, of impact and vacuum were out of the way. There were no clear new laws of motion yet; it was left to Newton to set these out; but there were fair descriptions of where and how masses in fact move, and no interest at all in where they ought to want to move.[1]

What was the nature of Newton's insight? How did he exercise those great gifts, and seize the great opportunity which I have described?

If we put what he did most baldly, it is this: that he carried on the simplification which Kepler had begun, but carried it beyond geometry into physics. Ptolemy, Copernicus, Tycho Brahe and Kepler, at bottom all looked no further than to plot the paths of the planets. Kepler found likenesses between these paths deeper than anything in the traditional astronomy, for his were likenesses of motion as well as shape. Nevertheless his paths remained descriptions, more accurate and more concise than Ptolemy's, but no more universal. For even when Kepler speculated about an attraction of the planets to the sun he had no principle to link it to the movement of earthly masses. Galileo had the first glimpse of that; and there were others as the seventeenth century marched on, who knew what kind of principle they were looking for; but it was Newton who formulated it, sudden and entire. He said that change of motion is produced by force; that the motion between masses, whether apple, moon and earth, or planet and sun, is produced by gravitational forces which attract them to one another. And he alone of his contemporaries had the mathematical power to show that, if these forces are postulated in the right way, then they keep the planets spinning like a clockwork, they keep the moon on its orbit, and the tides moving under the moon; and they hold the universe together. These achievements are so great that they out-top astronomy; and they are only a part of Newton's whole achievement. But more than the achievement, it is the thought within which deserves our

---

[1] *Aristotle was more interested in final than in efficient causes. (The final cause of the stone's hitting Goliath's forehead was David's wish to kill him; the efficient cause was the impetus given the stone by the movement of certain muscles in David's arm.) There had to be a motive, as it were, for a moving object. So, stones and rain (earth and water) fell in order to rejoin and take their rightful place in the two innermost spheres of the universe, whereas bubbles and flames (air and fire) rose in order to become part of the next two spheres. (Ed.)*

study. There is the searching conception of the universe as a machine; not a pattern but a clockwork. There is the conception of the moving forces within the machine: the single spring of action in gravitation. There is the brilliant compromise between the description of the astronomers and the First Cause of the theologians, in which Newton shaped once for all the notion of cause as it has remained ever since. Newton indeed has taken over just enough of the Aristotelean nature of things to make the world work by giving all matter a single nature – that it seeks to join with all other matter.

J. BRONOWSKI, *The Common Sense of Science*, 1951

How fair is it to argue that where Greek astronomy described Newton explained? Certainly Newton's account is more satisfactory, in that it incorporates, and establishes a connexion between, observations of movements of many different kinds besides planetary ones. Equally certainly it is a limited one in that, although it is completely based on the idea of gravity, Newton 'evades the hypnosis of the insoluble problem'[1] and refuses to put forward any hypothesis as to the nature of this force which can operate across or through such empty distances. Does his use of this concept of gravity amount then, in the last analysis, to an explanation, or merely a very much more detailed description (a physical or mechanical description, if you will, rather than a geometric one) of the behaviour not just of planets but of apples and galaxies as well? When, in fact, do descriptions cease to be descriptions and become explanations?

As for whether Copernicus and Galileo were right in maintaining that the earth goes round the sun . . .

III. Relatively Speaking

Before Copernicus, people thought that the earth stood still and the heavens revolved about it once a day. Copernicus taught that 'really' the earth rotates once a day, and the daily revolution of

[1] *H. T. Pledge*, Science Since 1500, *London 1939.*

sun and stars is only 'apparent'. Galileo and Newton endorsed this view, and many things were thought to prove it – for example the flattening of the earth at the poles, and the fact that bodies are heavier there than at the equator. But in the modern theory the question between Copernicus and his predecessors is merely one of convenience; all motion is relative, and there is no difference between the two statements: 'the earth rotates once a day' and 'the heavens revolve about the earth once a day'. The two mean exactly the same thing, just as it means the same thing if I say that a certain length is six feet or two yards. Astronomy is easier if we take the sun as fixed than if we take the earth, just as accounts are easier in a decimal coinage. But to say more for Copernicus is to assume absolute motion, which is a fiction. All motion is relative, and it is a mere convention to take one body as at rest. All such conventions are equally legitimate, though not all are equally convenient.

BERTRAND RUSSELL, *The ABC of Relativity*, 1925

The next passage, though perhaps conceding premature victory to one of the two sides to an argument which still goes on, is an admirable summary of much recent astronomic thought.

It also gives a good idea of how scientists are sometimes able to attack a problem as seemingly insoluble as how the universe began, and of the peculiar difficulties which face astronomers more than most scientists. For, although geologists and evolutionary biologists, for instance, are also obliged to speculate about events which happened long ago and which may in many respects have been unlike anything happening now, and although, at the other end of the scale, the behaviour of electrons or the complex and all-important internal architecture of protein molecules must seem as exasperatingly inaccessible to the clumsy and inadequate means of perception at man's disposal as is the nature of those galaxies we can only hear, and at that faintly across vast stretches of both time and space, the obstacles to an advance of astronomical knowledge are still probably uniquely intimidating.

IV. The Origin of the Universe

Just as it was natural to think of the earth as the centre of the solar system, so for a long time there was a tendency to think of the sun as the centre of the starry system. But during this century the sun has been steadily demoted. It is now seen as a rather modest star situated towards the edge of a disc-shaped galaxy or array of stars. This galaxy is itself one member of a 'local' cluster of galaxies – and thousands of such clusters are scattered through the universe.

At the same time, our scale of cosmic distance has stretched vertiginously. By the turn of the century, the nearest star was known to be four 'light years' away, or 2,352,000 million miles. By 1930 it was agreed that the diameter of our galaxy is about 100,000 light years, while a neighbouring galaxy in Andromeda is nearly one million light years away. By 1936 the American, Edwin Hubble, was speaking of galaxies 500 million light years away, but in 1952 Walter Baade reinterpreted the observations and blithely multiplied the size of Hubble's universe by two. Subsequent corrections have again tripled this, and distances of several thousand million light years are now almost commonplace.

But the most startling discovery has been that all these distant galaxies appear to be hurtling away from us – and the farther away they are, the faster they hurtle. This is the starting point for all modern cosmologies.

The observation was based on a phenomenon which may be experienced on a railway platform. The whistle of an approaching express train sounds higher than that of a receding one – the pitch drops suddenly as the train roars through the station. This is because the sound waves are crowded together, so to speak, as the train approaches, and stretched out as it recedes.

The same effect applies to light waves. An approaching galaxy, blowing its luminous train whistle, should shift its light to a higher 'pitch' – it looks slightly more blue in colour. A receding galaxy looks redder. The effect is very slight, but it can be detected by measuring a shift in certain dark lines which appear when the light from the galaxy is analysed with a spectroscope.

By 1930 it had become clear that most of the galaxies showed a 'red shift', and were thus receding. But even more startling, the farther the galaxy, the greater the shift. Last summer, an object was photographed whose red shift is so great that it must be receding at nearly half the speed of light, and is about 4,500 million light years away.

If this relation between speed and distance holds still farther out, there could be galaxies receding from us so fast that their light never reaches us, and they can never be observed. Thus the speed of light now defines the ultimate frontier of the observable universe. It is within this frontier that the cosmologist must go to work.

The discovery of the receding galaxies led straight to the 'big bang' theory of the universe. If the galaxies are getting farther apart, they must originally have been closer together. By measuring the speed of expansion and reversing it, one can calculate a time in the past when all the galaxies were packed close together.

At this time, it was supposed, the universe may have consisted of a 'primeval atom' of incredibly concentrated energy. This atom then blew apart, and the universe has been expanding like a bomb ever since. In the future, it may simply disperse and run down like a clock – or it might start to contract again into another primeval atom.

Some scientists found this picture profoundly unsatisfactory. It raises unanswerable questions about the origin of the universe, the nature of the primeval atom, or the cause of the big bang. Why not simply abolish these questions?

The 'steady state' theory, first discussed half as a joke by Bondi and Gold, asks us to imagine a universe which has always existed and always will, and does not evolve in any way. It would look broadly the same anywhere in space and anywhere in time. This implies that the average number of galaxies in any part of space must stay the same. But astronomy shows that the galaxies are all rushing away from each other. Therefore, said the steady state theory, we must postulate a continuous creation of matter to fill up the gaps.

This, it was suggested, simply appears continuously in empty space in the form of hydrogen. It condenses to form new galaxies, which develop just fast enough to balance the dispersion of existing ones, thus maintaining the average galaxy population for every part of the universe at the same time. Fred Hoyle showed that this daring idea, apparently contradicting classical laws of physics about the conservation of matter and energy, could be firmly based on a development of the relativity theory.

These ideas have gained momentum. Their great attraction was that they could be tested by experiment. If the universe is evolving, the galaxies would have been closer together in the past than they are to-day; but if the universe is in a steady state, it would look the same however far back in time we went.

Now astronomy is in the curious situation that it can undertake a kind of time travel and test these alternatives directly. The reason is that the stars and galaxies are so far away that light takes an enormous time to reach us. Light reaching us now left the more distant galaxies thousands of millions of years ago. We are thus not seeing them *as they are now*, but *as they were*. What is more, the stars that are farthest off in space are farthest off in time. Probing out into space with telescopes is like digging down a geological stratum of fossilised light. And a study of the most distant galaxies can indicate what the universe was like long ago.

If the big bang theory is right, the most distant galaxies we see should be closer together – since we are seeing them as they were nearer to the time of the bang. But if the 'steady state' theory is right, the distant galaxies should be distributed in the same way as the near ones. This was the most direct and obvious test suggested by the theory.

But a snag cropped up. It turned out that optical telescopes could not hope to see quite far enough out into space (or, if you prefer, back into time). The 200-inch telescope at Mount Palomar, California, the largest in the world, can just begin to probe the regions where there should be a detectable difference between an evolving and a 'steady state' universe. But the results are too marginal to be conclusive.

At this point, radio astronomy came into its own. Surveys of the sky with radio telescopes revealed a considerable number of 'radio stars'. These were point-like sources of radio waves in the sky, only a few of which seemed associated with objects visible through telescopes. But in 1952, Walter Baade turned the 200-inch telescope on the position of an intense radio star in the constellation of Cygnus which had been accurately pin-pointed by F. G. Smith at Cambridge. He succeeded in photographing a most peculiar object; it appeared to be a pair of galaxies in collision about 500 million light years away. Another theory holds that it is one galaxy splitting in two, but the important thing is that it emits extremely powerful radio waves. Optically it is almost invisible, but to the radio telescope it is as 'bright' as the sun.

It was immediately realised that if such a distant object could generate such powerful radio waves, similar objects at far greater distances would still be detectable by radio telescopes. In other words, the radio astronomers could probe much further into space and time, and settle the cosmological controversy.

This, in effect, is what the Cambridge team under Professor Martin Ryle are now confident that they have done. Since 1958, the team have been using a new and extremely powerful radio telescope. With this, they have made the most detailed survey and analysis of parts of the radio sky yet undertaken. On it, they have based the elaborate studies presented to the Royal Astronomical Society last week. The work is intricate, and the argument involved – but the gist of it is this:

If, like the radio telescope, we could perceive the radio waves from the sky, we would see many bright radio stars on a dimly luminous background of general radiation. We would ask, straight away: what are the radio stars, how far away are they, and what causes the background?

The first move was to direct optical telescopes at radio stars, and see if they are associated with any visible objects. Most of them, it turns out, are not. A few radio stars have been identified with visible objects inside our galaxy; a few more represent radio emission from nearby galaxies; and a rather larger number have been identified with extremely faint objects, some of which are probably colliding (or splitting) galaxies like the Cygnus source. These are a million times more powerful radio emitters than our own or nearby galaxies.

But there remain a very large number of radio stars for which no corresponding visible objects can be found. They could be nearby objects, which for some reason give out radio waves but little or no light. Or they could be very distant objects like the Cygnus source – in which case they might be used to settle the cosmological question. How are the two possibilities to be distinguished?

Suppose first, the Cambridge team said, that most of the radio stars are not far away, but are associated with our own galaxy. Suppose, too, that they are all powerful enough to be detected by the radio telescope. In this case, since we are situated towards the edge of our galaxy, we would expect to see more radio stars towards the galactic centre than towards the edge.

One Cambridge study, therefore, was to analyse the distribution of the radio stars in the sky – and it was shown conclusively that they are distributed evenly.

Does this prove that the radio stars are beyond the galaxy? No – there is another possibility: they might be very weak radio emitters, so that the radio telescope detects only the nearest ones, *well inside* the galaxy. These would then represent a small sample of the total radio star population of the galaxy. The rest of this population, too faint to be distinguished individually, would contribute to the general luminous background.

If we assume the detectable radio stars are all very weak, they must all be very near, and hence packed rather densely round us. But the more distant, undetectable ones must be equally closely packed – so the galaxy as a whole must be very densely populated with these weak, radio emitting objects. This large population would contribute to the luminous radio background – and would make it very bright. In fact, measurements of the background show that it is nowhere near bright enough, and it turns out that the only way of reconciling the observed background with the observed number of radio stars is to suppose that most of these stars are *outside* the galaxy.

This argument was then extended beyond our galaxy, and it was proved that most radio stars are not only outside the galaxy, but are very far away. The majority of them, in fact, must be rare, very powerful sources, comparable to the Cygnus source, but much more distant. The sources now being detected at Cambridge, in fact, are providing a picture of the universe as it was some 8,000 million years ago – and the sources themselves are receding at something approaching nine-tenths of the speed of light.

The final step at Cambridge was to work out the cosmological implications of this. On the big bang theory, the radio star population should get denser as you go back in time – there should be more of the faintest, most distant sources. On the 'steady state' theory, the population density should stay the same. The Cambridge observations have shown now unmistakably that the density of weaker sources is at least three times, and probably ten times, higher than the 'steady state' theory predicts. Therefore, the universe must be evolving.

JOHN DAVY, from an article in the *Observer*, 12 February 1961[1]

---

[1] *The* Sunday Times *of the same date reported Professor Fred Hoyle as being 'more confident than ever' in the theory of continuous creation. One possible alternative interpretation of the Cambridge results, according to Hoyle, was that a higher proportion of the very old, or very distant, galaxies might well be of the peculiar type which acts as a powerful source of radio waves. However, by 1964 he was postulating as a more likely source of these radio waves the immensely distant super stars, quasi-stars or quasars as they are variously called, which are central to his latest theories. These are condensations of cosmic dust such as would, in the normal run, break up into separate blobs or stars and form galaxies, but which for some reason or other continue to contract as a whole, and so form gigantic single stars which are imploding under the enormous force of gravity they generate. These implosions might, thinks Hoyle, squeeze out great jets of matter which could be the immensely powerful sources of radio waves detected by radio telescopes. (Ed.)*

CHAPTER TWO
The Nature of Matter

I. Atoms

His (Democritus') opinions are these. The first principles of the universe are atoms and empty space; everything else is merely thought to exist. The worlds are unlimited; they come into being and perish. Nothing can come into being from that which is not nor pass away into that which is not. Further, the atoms are unlimited in size and number, and they are borne along in the whole universe in a vortex, and thereby generate all composite things – fire, water, air, earth; for even these are conglomerations of given atoms. And it is because of their solidity that these atoms are impassive and unalterable. The sun and the moon have been composed of such smooth and spherical masses, i.e. atoms, and so also the soul, which is identical with reason. We see by virtue of the impact of images upon our eyes.

All things happen by virtue of necessity, the vortex being the cause of the creation of all things, and this he calls necessity. The end of action is tranquility, which is not identical with pleasure, as some by false interpretation have understood, but a state in which the soul continues calm and strong, undisturbed by any fears or superstition or any other emotion. This he calls well-being, and many other names. The qualities of things exist merely by convention; in nature there is nothing but atoms and void space. These, then, are his opinions.

DIOGENES LAERTIUS (summarizing the views of DEMO-CRITUS, *circa* 460–370 B.C.) (tr. R. D. Hicks)

II. The Four Elements

Of the first philosophers, most thought the principles which were of the nature of matter were the only principles of all things;[1]

[1] *Aristotle himself is also concerned with the* form *of things and how this comes into being, as well as the mere substance out of which they are formed. (Ed.)*

that of which all things that are consist, and from which they first come to be, and into which they are finally resolved (the substance remaining, but changing in its modifications), this they say is the element and principle of things, and therefore they think nothing is either generated or destroyed, since this sort of entity is always conserved, as we say Socrates neither comes to be absolutely when he comes to be beautiful or musical, nor ceases to be when he loses these characteristics, because the substratum, Socrates himself, remains. So they say nothing else comes to be or ceases to be; for there must be some entity – either one or more than one – from which all other things come to be, it being conserved.

Yet they do not all agree as to the number and the nature of these principles. Thales, the founder of this school of philosophy, says the principle is water (for which reason he declared that the earth rests on water), getting the notion perhaps from seeing that the nutriment of all things is moist . . .

Anaximines and Diogenes make air prior to water, and the most primary of the simple bodies, while Hippasus of Metapontium and Heraclitus of Ephesus say this of fire, and Empedocles says it of the four elements, adding a fourth – earth – to those which have been named; for these, he says, always remain and do not come to be, except that they come to be more or fewer, being aggregated into one and segregated out of one.

ARISTOTLE (384–322 B.C.), *Metaphysica* (tr. W. D. Ross)

The aim since very early, then, has been to get as near as possible to a single substance, element, continuum – what you will – capable of diversifying itself into all those materials out of which the sensible universe seems to be constructed. Of almost equal antiquity are the atomic hypothesis and the theory of the four elements. Scientifically the former seemed to answer no more, perhaps even fewer, questions than the latter ; philosophically it was used as the basis of a disturbingly impious, materialistic determinism. (Lucretius, a Roman disciple of Democritus, gives a fuller statement of this in his De Rerum Natura.) *As for verification, so far as such a notion was understood or thought necessary, common sense seemed to do it for the four elements quite adequately.*

Not till the Renaissance and after was the idea that the universe was made up of atoms seriously revived, this time not as a piece of philosophical speculation but as a means of accounting for certain everyday natural phenomena. And first, as put so graphically by Hooke in the next passage, atoms can be used to explain something of the nature of heat and its link with the three states of matter. Hooke is also trying to use atoms chemically, searching for some affinity, likeness or congruity between atoms of different substances which would lead to the willingness or unwillingness of those substances to unite. This is yet another instance of the versatile Hooke being on the right track (elsewhere in Micrographia, *for example, he comes very close to explaining the true nature of combustion some hundred years too early), but it is left to Dalton to develop the idea to a point where it bears fruit.*

Newton is a physicist, and in the Principia *he certainly uses the concept of atoms physically, to enable him to think of stars and other spherical agglomerations of matter cohering in space, held together like raindrops but by the force of gravity. Here, however, he is either very close to speculating for speculating's sake, as he so wisely refused to do in the case of gravity, about the* nature *of atoms, or he is searching for those qualities in atoms which will explain the stability, in a world of flux, of* chemical *properties.*

Dalton, being clearer in his mind than Hooke about the distinction between a mixture and a compound, and knowing that in the latter case the proportions of the two substances remain constant, took the further step of postulating that such combination of substances took place on an atom-to-atom basis, whether one to one, one to two, or so forth, and that the relative numbers of atoms of the substances going to make up each compound particle (or molecule) also remain constant. Hence, if such numbers can be inferred, and if the relative weights 'in the mass' of the substances involved be determined accurately, the relative weights of their respective atoms can be worked out.

Note the trend in prose style. Is the gap of forty years sufficient to

account for the difference between Hooke and Newton in this respect,
or do the styles reflect the men – the one with a hundred bright ideas,
the other with a single great idea?

III. Like Grains of Sand

And that we may the better find out what the cause of *Congruity*
and *Incongruity*[1] in bodies is, it will be requisite to consider, First,
what is the *cause* of *fluidness*; And this, *I conceive*, to be nothing
else but a certain *pulse* or *shake* of *heat*; for Heat being nothing else
but a very *brisk* and *vehement agitation* of the parts of a body, (as
I have elsewhere made *probable*) the parts of a body are thereby
made so *loose* from one another, that they easily *move any way*,
and become *fluid*. That I may explain this a little by a gross
Similitude, let us suppose a dish of sand set upon some body that
is very much *agitated*, and shaken with some *quick* and *strong*
vibrating motion, as on a *Milstone* turn'd round upon the under
stone very violently whilst it is empty; or on a very stiff *Drum*-
head, which is vehemently or very nimbly beaten with the Drum-
sticks. By this means, the sand in the dish, which before lay like a
dull and unactive body, becomes a perfect *fluid*; and ye can no
sooner make a *hole* in it with your finger, but it is immediately
filled up again, and the upper surface of it *levell'd*. Nor can you
bury a *light body*, as a piece of Cork under it, but it presently
emerges or *swims* as 'twere on the top; nor can you lay a *heavier*
on the top of it, as a piece of Lead, but it is immediately *buried* in
Sand, and (as 'twere) sinks to the bottom. Nor can you make a
hole in the side of the Dish, but the sand shall *run out* of it to a
level.[2] Not an *obvious property* of a fluid body, as such, but this
dos *imitate*; and all this meerly caused by the vehement *agitation*
of the conteining vessel; for by this means, *each* sand becomes to
have a *vibrative* or *dancing* motion, so as no other heavier body
can *rest* on it, unless *sustein'd* by some other on either side: Nor

[1] *Hooke is concerned as to why certain fluids will mix, and others, like oil and*
water, will not. (Ed.)

[2] *Comma in the original.* (Ed.)

will it suffer any Body to be *beneath* it, unless it be a *heavier* then it self . . .

Having therefore in short set down my Notion of a Fluid body, I come in the next place to consider what *Congruity* is; and this, as I said before, being a *Relative property* of a fluid, whereby it may be said to be *like* or *unlike* to this or that other body, whereby it *does* or *does not mix* with this or that body. We will again have recourse to our former Experiment, though but a rude one; and here if we mix in the dish *several kinds* of sands, some of *bigger*, others of *less* and finer bulks, we shall find that by the agitation the *fine sand* will *eject* and *throw out* of it self all those *bigger* bulks of small *stones* and the like, and those will be *gathered* together all into *one* place; and if there be *other* bodies in it of other natures, those also will be *separated* into a place by themselves, and *united* or *tumbled* up together. And though this do not come up to the *highest property* of Congruity, which is a *Cohæsion* of the parts of the fluid together, or a kind of *attraction* and *tenacity*, yet this does as 'twere *shadow* it out, and somewhat resemble it; for just after the same manner, I suppose the *pulse* of heat to *agitate* the small parcels of matter, and those that are of a *like bigness*, and *figure*, and *matter*, will *hold*, or *dance* together, and those which are of a *differing* kind will be *thrust* or *shov'd* out from between them. . . .

ROBERT HOOKE, Micrographia, 1665

IV. Solid, Massy and Hard

All these things being consider'd, it seems probable to me, that God in the Beginning form'd Matter in solid, massy, hard, impenetrable, moveable Particles, of such Sizes and Figures, and with such other Properties, and in such Proportion to Space, as most conduced to the End for which he form'd them; and that these primitive Particles being Solids, are incomparably harder than any porous Bodies compounded of them; even so very hard, as never to wear or break in pieces; no ordinary Power being able to divide what God himself made one in the first Creation. While the Particles continue entire, they may compose Bodies of one and the same Nature and Texture in all Ages: But should they wear away, or break in pieces, the Nature of Things depending on them, would be changed. Water and Earth, composed of old worn Particles and Fragments of Particles, would not be of the same Nature and Texture now, with Water and Earth composed of entire Particles in the Beginning. And therefore, that Nature may be lasting, the Changes of corporeal Things are to be placed only in the various Separations and new Associations and Motions of these permanent Particles; compound Bodies being apt to break, not in the midst of solid Particles, but where those Particles are laid together, and only touch in a few Points.

ISAAC NEWTON, *Opticks*, 1704 (passage taken from fourth edition, corrected, 1730)

V. Atoms and Chemistry

[*a*] There are three distinctions in the kinds of bodies, or three states, which have more especially claimed the attention of philosophical chemists; namely, those which are marked by the terms *elastic fluids, liquids, and solids.* A very familiar instance is exhibited to us in water, of a body, which, in certain circumstances, is capable of assuming all the three states. In steam we recognise a perfectly elastic fluid, in water, a perfect liquid, and in ice a complete solid. These observations have tacitly led to the conclusion which seems universally adopted, that all bodies of sensible magnitude, whether liquid or solid, are constituted of a vast number of extremely small particles, or atoms of matter bound together by a force of attraction, which is more or less powerful according to circumstances. . . .

Whether the ultimate particles of a body, such as water, are all alike, that is, of the same figure, weight, &c. is a question of some importance. From what is known we have no reason to apprehend a diversity in these particulars: if it does exist in water, it must equally exist in the elements constituting water, namely, hydrogen and oxygen. Now it is scarcely possible to conceive how the aggregates of dissimilar particles should be so uniformly the same. If some of the particles of water were heavier than others, if a parcel of the liquid on any occasion were constituted principally of these heavier particles, it must be supposed to affect the specific gravity of the mass, a circumstance not known. Similar observations may be made on other substances. Therefore we may conclude that *the ultimate particles of all homogeneous bodies are perfectly alike in weight, figure*, &c. In other words, every particle of water is like every other particle of water; every particle of hydrogen is like every other particle of hydrogen, &c.

[*b*] When any body exists in the elastic state, its ultimate particles are separated from each other to a much greater distance than in any other state; each particle occupies the centre of a comparatively large sphere, and supports its dignity by keeping all the rest, which by their gravity, or otherwise are disposed to encroach up(on) it, at a respectful distance. When we attempt to conceive the *number* of particles in an atmosphere, it is somewhat like attempting to conceive the number of stars in the universe; we are confounded with the thought. But if we limit the subject, by taking a given volume of any gas, we seem persuaded that, let the divisions be ever so minute, the number of particles must be finite; just as in a given space of the universe, the number of stars and planets cannot be infinite.

Chemical analysis and synthesis go no farther than to the separation of particles one from another, and to their reunion. No new creation or destruction of matter is within the reach of chemical agency. We might as well try to introduce a new planet into the solar system, or to annihilate one already there in existence, as to destroy or create a particle of hydrogen. All the changes we can produce, consist in separating particles that are in a state of cohesion or combination, and joining those that were previously at a distance.

In all chemical investigations, it has justly been considered an important object to ascertain the relative *weights* of the simples which constitute the compound. But unfortunately the enquiry has terminated here; whereas from the relative weights in the mass, the relative weights of the ultimate particles or atoms of the bodies might have been inferred, from which their number and weight in various other compounds would appear, in order to assist and guide future investigations, and to correct their results. Now it is one great object of this work, to shew the importance and advantage of ascertaining *the relative weights of the ultimate particles, both of simple and compound bodies, the number of simple elementary particles which constitute one compound particle, and the number of less compound particles which enter into the formation of one more compound particle.*

If there are two bodies, A and B, which are disposed to combine, the following is the order in which the combinations may take place, beginning with the most simple: namely,

> 1 atom of A + 1 atom of B = 1 atom of C, binary
> 1 atom of A + 2 atoms of B = 1 atom of D, ternary
> 2 atoms of A + 1 atom of B = 1 atom of E, ternary
> 1 atom of A + 3 atoms of B = 1 atom of F, quaternary
> 3 atoms of A + 1 atom of B = 1 atom of G, quaternary
> &c. &c.

The following general rules may be adopted as guides in all our investigations respecting chemical synthesis.

1st. When only one combination of two bodies can be obtained, it must be presumed to be a *binary* one, unless some cause appear to the contrary.

2nd. When two combinations are observed, they must be presumed to be a *binary* and a *ternary*.

3rd. When three combinations are obtained, we may expect one to be a *binary*, and the other two *ternary*.

4th. When four combinations are observed, we should expect one *binary*, two *ternary*, and one *quaternary*, &c.

5th. A *binary* compound should always be specifically heavier than a mere mixture of its two ingredients.

6th. A *ternary* compound should be specifically heavier than the mixture of a binary and a simple, which would, if combined, constitute it; &c.

7th. The above rules and observations equally apply, when two bodies, such as C and D, D and E, &c. are combined.

From the application of these rules, to the chemical facts already well ascertained, we deduce the following conclusions; 1st. That water is a binary compound of hydrogen and oxygen, and the relative weights of the two elementary atoms are as 1 : 7, nearly;[1] 2nd. That ammonia is a binary compound of hydrogen and azote,[2] and the relative weights of the two atoms are 1 : 5, nearly; 3rd. That nitrous gas is a binary compound of azote and oxygen, the atoms of which weigh 5 and 7 respectively; that nitric acid is a binary or ternary compound according as it is derived, and consists of one atom of azote and two of oxygen, together weighing 19; that nitrous oxide is a compound similar to nitric acid, and consists of one atom of oxygen and two of azote, weighing 17; that nitrous acid is a binary compound of nitric acid and nitrous gas, weighing 31; that oxynitric acid is a binary compound of nitric acid and oxygen, weighing 26; 4th. That carbonic oxide is a binary compound, consisting of one atom of charcoal[3] and one of oxygen, together weighing nearly 12; that carbonic acid is a ternary compound (but sometimes binary) consisting of one atom of charcoal, and two of oxygen, weighing 19; &c. &c. In all these cases the weights are expressed in atoms of hydrogen,[4] each of which is denoted by unity.

JOHN DALTON, *A New System of Chemical Philosophy*, 1808

---

[1] *Dalton's detailed conclusions are not all to be trusted: his experimental apparatus was often not accurate enough (he himself later amended the Hydrogen-Oxygen ratio to 1 : 8), and in this case he mistook a ternary compound (H_2O) for a binary. (Ed.)*

[2] *Nitrogen. (Ed.)*

[3] *Carbon. (Ed.)*

[4] *Dalton took the weight of one hydrogen atom as the unit in terms of which to express the atomic weights of all other elements. For a while these worked out at so*

Dalton is often described as the father of modern chemistry. And certainly his linking together of atoms and chemistry led directly to

nearly whole numbers that his contemporary Prout was able to suggest that all other atoms were built up of hydrogen atoms. Then certain elements (chlorine for example) were discovered to have atomic weights which were far from whole numbers. The problem was not resolved until more was known of the structure of the atom.

As Heisenberg explains in the next passage, the atom is made up of protons and neutrons tightly bound up together in a nucleus, and electrons circling this nucleus at a relatively great distance. Each proton carries a positive electrical charge, each electron an equivalent negative charge. Neutrons are electrically neutral, and, as an atom normally has equal numbers of protons and electrons, so is the whole atom. It is these electrically charged particles which determine an atom's chemical properties, since chemical combination consists of a complex interaction between the outer electrons of two or more atoms, and whether or not this can take place depends on the number and disposition of those electrons. Nowadays an element's atomic number, which is determined by the number of protons in its nucleus (and hence in most cases of electrons circling round), and which has a direct bearing on its chemical properties, is considered of more importance than its atomic weight.

The weight of an atom, on the other hand, depends almost exclusively on its protons and neutrons, which are of practically the same mass as each other, while electrons are of negligible mass. Hence it is possible for two or more atoms to have identical numbers of protons and electrons, and therefore to behave indistinguishably from one another in a purely chemical sense, and yet to have differing numbers of neutrons bound up with the protons in the nucleus, and so to have differing atomic weights. These separate forms of the same chemical element are known as isotopes. Each isotope, when considered in isolation, has an atomic weight which is a whole number, since the components which give it weight (protons and neutrons) are each of the same mass as the single proton which forms the nucleus of a hydrogen atom. But in nature the atoms of different isotopes within one element are always found mixed up together (indeed, as the makers of the first atom bombs discovered, it is difficult to isolate an appreciable quantity of a particular isotope), so the atomic weight of an element whose atoms are found in two or more isotope forms is always an average of the atomic weights of those isotopes.

Note that Prout's hypothesis, or guess, even though it ran into evidence which seemed to rule it out, was fairly near the mark in the end. (As Conant says on page 81, 'A theory is only overthrown by a better theory, never merely by contradictory facts.') Note also that Dalton, unlike Prout, did not concern himself with more philosophical speculations as to the ultimate nature of atoms, but merely with hypotheses he could use. (Ed.)

most of the real advances in that science during the nineteenth century.

Towards its close, however, and certainly in the twentieth century, atoms have reverted to being the concern, in the main, of the physicist. And although his discoveries have had remarkably and even terrifyingly practical results, his concern when carrying out the original pure research has almost always been to discover something quite useless about the ultimate nature of atoms and of matter.

VI. Atomic Particles

At the beginning of the modern age the atomic concept was closely linked with that of chemical elements. An element was characterized by the fact that it could not be decomposed further by chemical means, and by the fact that a particular kind of atom belonged to a particular element. Thus pure carbon consists of carbon atoms only, and pure iron of iron atoms only, and one was forced to assume that there were just as many sorts of atoms as there were chemical elements. Since ninety-two different kinds of elements were eventually known, one assumed the existence of ninety-two kinds of atoms.

However, this conception is not satisfying if we approach the problem from the basic premises of the atomic theory. Originally atoms were introduced to explain matter qualitatively through their movements and structure. This conception can have a true explanatory value only if all the atoms are equal, or if there are but a few kinds of atoms – in other words, if atoms themselves have no qualities at all. . . . Quite early on it was thought that chemical atoms themselves might well be composed of a very small number of basic building-stones. After all, even the oldest attempts to change one chemical substance into another must have been based on the assumption that in the final analysis all matter was one.

The last fifty years have shown that all chemical atoms are composed of only three basic building-stones, which we call protons, neutrons and electrons. The atomic nucleus consists of protons and neutrons and is surrounded by a number of electrons. Thus, for instance, the nucleus of the carbon atom consists of six protons and six neutrons and is surrounded by six electrons at a relatively great distance. Thanks to the development of nuclear physics in the 1930's we now have these three different kinds of particles, instead of the ninety-two different kinds of atoms, and in this respect atomic theory has followed the very path which its basic assumptions had suggested. When it became clear that all the atoms were composed of only three kinds of basic building-stones, there arose the practical possibility of changing chemical elements into one another. We know that this physical possibility was soon to be followed by its technical realization. Since Otto Hahn's *discovery of the fission of uranium* in 1938, and following the technical developments which sprang from it, we can now transform elements even on a large scale.

However, during the last two decades the picture has once more become a little confused. In addition to the three elementary particles already mentioned, the proton, neutron and electron, new elementary particles were discovered in the thirties and during the last few years their number has increased most disturbingly. In contrast to the three basic building-stones, these new particles are always unstable and have very short lifetimes. Of these so-called *mesons*, one type has a lifetime of about a millionth of a second, another lives for only a hundredth part of that time, and a third, which has no electrical charge, for only a hundred billionth of a second. However, apart from their instability, these three new elementary particles behave very similarly to the three stable building-stones of matter. At first glance it looks as if, once more, we were forced to assume the existence of a greater number of qualitatively different particles, which would be most unsatisfactory in view of the basic assumptions of atomic physics. However, during experiments in the last few years, it has become clear that these elementary particles can change into one another during their collisions, with great changes in energy. When two elementary particles collide with a great energy of motion, new elementary particles are created and the original particles, together with their energy, are changed into new matter.

This state of affairs is best described by saying that all particles are basically nothing but different stationary states of one and the same stuff. Thus even the three basic building-stones have become reduced to a single one. *There is only one kind of matter but it can exist in different discrete stationary conditions.* Some of these conditions, i.e. protons, neutrons and electrons, are stable while many others are unstable.

WERNER HEISENBERG, *The Physicist's Conception of Nature*, 1958 (tr. Arnold J. Pomerans: first published in German 1955)

VII. Energy, Matter and Light

Up till now we have often opposed light and matter, the latter being associated with mass and often with electricity, while the former always appears free from inertia and charge. But if these two fundamental entities of the physical world appear to oppose one another, they are none the less related because they are both special forms of energy. In principle, therefore, there is nothing against the view that energy, while always conserving itself, can pass from the material to the luminous form and vice versa. We know today that it is actually so; this fact breaks down the barrier which seemed to separate light and matter and, to complete the enumeration of the fundamental properties which assure light a privileged place among the physical entities, we can now add that *light is, in short, the most refined form of matter.* . . .

Giving free scope to our imagination, we could suppose that at the beginning of time, on the morrow of some divine 'Fiat Lux', light, at first alone in the universe, has little by little produced by progressive condensation the material universe such as, thanks to light itself, we can contemplate it today. And perhaps one day, when time will have ended, the universe, recovering its original purity, will again dissolve into light.

LOUIS DE BROGLIE, *Physics and Microphysics*, 1955 (tr. Martin Davidson: first published in French 1947)

The wheel has come full circle. 'There is only one kind of matter, but it can exist in different discrete stationary conditions,' says Heisenberg; and Aristotle: 'There must be some entity – either one or more than one – from which all other things come to be, it being conserved.' Moreover, 'We see by virtue of the impact of images upon our eyes', since 'Light is, in short, the most refined form of matter.' And if, in fact, 'there is nothing against the view that energy, while always conserving itself, can pass from the material to the luminous form and vice versa', who is to smile at 'The sun and moon have been composed of such smooth and spherical masses, i.e. atoms, and so also the soul, which is identical with reason'? If all is not material, all may yet be immaterial.

CHAPTER THREE

The Nature of Life

Not even a pretence has been made in this chapter to present, as in the previous two, a summary of changing views on the subject in question. This is largely because I have found it impossible, in a short extract, to begin to do justice to all the most recent and exciting work on the chemistry of the cell. For those interested, however, there are suggestions for further reading at the end of the book.

Instead there are examples of three ways of thinking about just certain aspects to the nature of life – three ways of thinking which are also one way of thinking.

The first and third extracts are both from public lectures: the first with much of the eloquence and some of the condescension of a nineteenth-century platform manner, the other with the intimacy and informality of a twentieth-century radio manner.

1. Machines

A horse is not a mere dead structure: it is an active, living, working machine. Hitherto we have, as it were, been looking at a steam-engine with the fires out, and nothing in the boiler; but the body of the living animal is a beautifully formed active machine, and every part has its different work to do in the working of that machine, which is what we call its life. The horse, if you see him after his day's work is done, is cropping the grass in the fields, as it may be, or munching the oats in his stable. What is he doing? His jaws are working as a mill – and a very complex mill too – grinding the corn, or crushing the grass to a pulp. As soon as that operation has taken place, the food is passed down to the stomach, and there it is mixed with the chemical fluid called the gastric juice, a substance which has the peculiar property of making soluble and dissolving out the nutritious matter in the grass, and leaving behind those parts which are not nutritious; so that you have, first, the mill, then a sort of chemical digester; and then the food, thus partially dissolved, is carried back by the muscular contractions of the intestines into the hinder parts of the body, while the soluble portions are taken up into the blood. The blood is contained in a vast system of pipes, spreading through the whole body, connected with a force-pump, – the heart, – which, by its position and by the contractions of its valves, keeps the blood constantly circulating in one direction, never allowing it to rest; and then, by means of this circulation of the blood, laden as it is with the products of digestion, the skin, the flesh, the hair, and every other part of the body, draws from it that which it wants, and every one of these organs derives those materials which are necessary to enable it to do its work.

T. H. HUXLEY, *Six Lectures to Working Men on the Phenomena of Organic Nature*, 1863

11. Machines and Whirlpools

Restricting our attention to the phenomena which have now been described, and to a short period in the life of the crayfish, the body of the animal may be regarded as a factory, provided with various bits of machinery, by means of which certain nitrogenous and other matters are extracted from the animal and vegetable substances which serve for food, are oxidated, and are then delivered out of the factory in the shape of carbonic acid gas, guanin, and probably some other products, with which we are at present unacquainted. And there is no doubt, that if the total amount of products given out could be accurately weighed against the total amount of materials taken in, the weight of the two would be found to be identical. To put the matter in its most general shape, the body of the crayfish is a sort of focus to which certain material particles converge, in which they move for a time, and from which they are afterwards expelled in new combinations. The parallel between a whirlpool in a stream and a living being, which has often been drawn, is as just as it is striking. The whirlpool is permanent, but the particles of water which constitute it are incessantly changing. Those which enter it, on the one side, are whirled around and temporarily constitute a part of its individuality; and as they leave it on the other side, their places are made good by new comers.

Those who have seen the wonderful whirlpool, three miles below the Falls of Niagra, will not have forgotten the heaped-up wave which tumbles and tosses, a very embodiment of restless energy, where the swift stream hurrying from the Falls is compelled to make a sudden turn towards Lake Ontario. However changeful in the contour of its crest, this wave has been visible, approximately in the same place, and with the same general form, for centuries past. Seen from a mile off, it would appear to be a stationary hillock of water. Viewed closely, it is a typical expression of the conflicting impulses generated by a swift rush of material particles.

Now, with all our appliances, we cannot get within a good many miles, so to speak, of the crayfish. If we could, we should see that it was nothing but the constant form of a similar turmoil of material molecules which are constantly flowing into the animal on the one side, and streaming out on the other.

> T. H. HUXLEY, *The Crayfish – An Introduction to the Study of Zoology*, 1880

III. Machines, Whirlpools and Populations

The model of the body as a machine has been the basis of much of the most useful part of our biology. With it we can find out how the body 'works', how much food must be supplied to provide fuel, and so on. In fact the machine model is the basis of much of agriculture and medicine . . . . (But) when we come to look more closely at the chemical interchanges between the body and the environment, we find that the body does not maintain a static structure, as the comparison with a machine suggests. Oxygen passes into the body, is burned, and is breathed out again as carbon dioxide. Was that oxygen ever part of the living body? Probably many people would say that it has merely passed through and been used by the body. But the paradox is that if one says this one will find that there is no enduring body at all. For we now know that even the apparently most stable parts, such as the bones, do not remain always compounded of the same actual molecules. If living things consist of no steady fabric of stuff but are continually changing, what is it then that is preserved? What is it about the chromosomes that makes them the bearers of heredity? What is it that makes each life in some sense the same from year to year? Individual chemical atoms remain in the cells for only a short time; what is preserved must be the pattern in which all these interchanging atoms are involved. 'Pattern,' you may say, 'what do you mean by that? Are you comparing me with a carpet?' That is just the trouble – we find it hard to find a proper model. Living patterns are not stable like those of pictures or carpets. A whirlpool might be a better simple analogy – the pattern of swirls in a river. The matter of these is continually changing and yet there is a sense in which each swirl remains the same. We might say that the flow through them is organized in a particular way. 'Yes, but organized by whom or what?' In the case of the river it is by the historical events of the past, which have left a certain arrangement, so that the water must flow through making just those patterns and swirls. So, in the case of the body, the organization is such that matter must flow through in certain ways.

Biology, like physics, has ceased to be materialist. Its basic unit is a non-material entity, namely an organization. But the organization is vastly more complicated than that of any river. It is kept in certain channels by the environment, acting in a sense as do the banks. If a stream stops, the banks remain, and therefore a river that has dried up may form again the same patterns. But the living patterns are so complicated that they are kept intact *only* by their continued activity. If they stop they are never re-started. The living patterns have developed a wonderful permanence none the less. They have the characteristic that every time there is any change in the banks the swirls make a compensating change and thus keep intact. The river analogy begins to fail us here and we may return instead to quite a different one.

If we want to understand an elaborate organization such as the body we may find it useful to compare it with an organization that we do understand. In an earlier lecture I compared the brain with a vast office devoted to keeping itself intact. I shall not press the analogy again here, but the development of the study of human organizations is beginning to provide us with powerful models for comparison with our own bodies. As sociology and the study of human organizations develop we are likely to find their terms useful in the very hard task of describing the living body, for which at present we have no proper model. The study of large populations and how they behave is one of the most recent human techniques. We now have methods of finding out what people do, by questionnaires, samples, market research, and other techniques, and statistics has a new set of symbols and methods of observation. These may help us greatly in studying the body, which is after all described as well by saying that it is a population of cells and chemical molecules as by comparing it with a clock. We can understand in this way how an organization endures although the separate individuals that make it up may change. Many people may laugh at this comparison. Application of a new model always seems absurd to those who use old ones. But it may be that we can find ways of controlling our ills better by speaking of ourselves as populations rather than as clocks.

J. Z. YOUNG, *Doubt and Certainty in Science*, 1951 (Reith Lectures, 1950)

CHAPTER FOUR

The Nature of Mind

1. Mind and Matter

I have settled down to the task of writing these lectures and have drawn up my chairs to my two tables. Two tables! yes; there are duplicates of every object about me – two tables, two chairs, two pens.

This is not a very profound beginning to a course which ought to reach transcendent levels of scientific philosophy. But we cannot touch bedrock immediately; we must scratch a bit at the surface of things first. And whenever I begin to scratch the first thing I strike is – my two tables.

One of them has been familiar to me from earliest years. It is a commonplace object of that environment which I call the world. How shall I describe it? It has extension; it is comparatively permanent; it is coloured; above all it is *substantial*. By substantial I do not merely mean that it does not collapse when I lean upon it; I mean that it is constituted of 'substance' and by that word I am trying to convey to you some conception of its intrinsic nature. It is a *thing*; not like space, which is a mere negation; nor like time, which is – Heaven knows what! But that will not help you to my meaning because it is the distinctive characteristic of a 'thing' to have this substantiality, and I do not think 'substantiality' can be described better than by saying that it is the kind of nature exemplified by an ordinary table. And so we go round in circles. After all, if you are a plain commonsense man, not too much worried with scientific scruples, you will be confident that you understand the nature of an ordinary table. I have even heard of plain men who had the idea that they could better understand the mystery of their own nature if scientists would discover a way of explaining it in terms of the easily comprehensible nature of a table.

Table No. 2 is my scientific table. It is a more recent acquaintance and I do not feel so familiar with it. It does not belong to the world previously mentioned – that world which spontaneously appears around me when I open my eyes, though how much of it is objective and how much subjective I do not here consider. It is part of a world which in more devious ways has forced itself on my attention. My scientific table is mostly emptiness. Sparsely scattered in that emptiness are numerous electrical charges rushing about with great speed, but their combined bulk amounts to less than a billionth of the bulk of the table itself. Notwithstanding its strange construction it turns out to be an entirely efficient table. It supports my writing paper as satisfactorily as table No. 1, for when I lay the paper on it the little electrical particles with their headlong speed keep on hitting the underside, so that the paper is maintained in shuttlecock fashion at a nearly steady level. If I lean upon this table I shall not go through; or, to be strictly accurate, the chance of my scientific elbow going through my scientific table is so excessively small that it can be neglected in practical life. Reviewing their properties one by one, there seems to be nothing to choose between the two tables for ordinary purposes; but when abnormal circumstances befall, then my scientific table shows to advantage. If the house catches fire my scientific table will dissolve quite naturally into scientific smoke, whereas my familiar table undergoes a metamorphosis of its substantial nature which I can only regard as miraculous. . . .

I need hardly tell you that modern physics has by delicate test and remorseless logic assured me that my second, scientific table is the only one which is really there – wherever 'there' may be. On the other hand I need not tell you that modern physics will never succeed in exorcising that first table – strange compound of external nature, mental imagery and inherited prejudice – which lies visible to my eyes and tangible to my grasp. We must bid good-bye to it for the present for we are about to turn from the familiar world to the scientific world revealed by physics. This is, or is intended to be, a wholly external world.

'You speak paradoxically of two worlds. Are they not really two aspects or two interpretations of one and the same world?'

Yes, no doubt they are ultimately to be identified after some fashion. But the process by which the external world of physics is transformed into a world of familiar acquaintance in human consciousness is outside the scope of physics. And so the world studied according to the methods of physics remains detached from the world familiar to consciousness, until after the physicist has finished his labours upon it. Provisionally, therefore, we regard the table which is the subject of physical research as altogether separate from the familiar world, without prejudicing the question of their ultimate identification. It is true that the whole scientific inquiry starts from the familiar world and in the end must return to the familiar world; but the part of the journey over which the physicist has charge is in foreign territory.

Until recently there was a much closer linkage; the physicist used to borrow the raw material of his world from the familiar world, but he does so no longer. His raw materials are æther, electrons, quanta, potentials, Hamiltonian functions, etc., and he is nowadays scrupulously careful to guard these from contamination by conceptions borrowed from the other world. There is a familiar table parallel to the scientific table, but there is no familiar electron, quantum or potential parallel to the scientific electron, quantum or potential. We do not even desire to manufacture a familiar counterpart to these things or, as we should commonly say, to 'explain' the electron. After the physicist has quite finished his world-building a linkage or identification is allowed; but premature attempts at linkage have been found to be entirely mischievous.

Science aims at constructing a world which shall be symbolic of the world of commonplace experience. It is not at all necessary that every individual symbol that is used should represent something in common experience or even something explicable in terms of common experience. The man in the street is always making this demand for concrete explanation of the things referred to in science; but of necessity he must be disappointed. It is like our experience in learning to read. That which is written in a book is symbolic of a story in real life. The whole intention of the book is that ultimately a reader will identify some symbol, say BREAD, with one of the conceptions of familiar life. But it is mischievous to attempt such identification prematurely, before the letters are strung into words and the words into sentences. The symbol A is not the counterpart of anything in familiar life. To the child the letter A would seem horribly abstract; so we give him a familiar conception along with it. 'A was an Archer who shot at a frog.' This tides over his immediate difficulty; but he cannot make serious progress with word-building so long as Archers, Butchers, Captains, dance round the letters. The letters are abstract, and sooner or later he has to realise it. In physics we have outgrown archer and apple-pie definitions of the fundamental symbols. To a request to explain what an electron really is supposed to be we can only answer, 'It is part of the ABC of physics'.

The external world of physics has thus become a world of shadows. In removing our illusions we have removed substance, for indeed we have seen that substance is one of the greatest of our illusions. Later perhaps we may inquire whether in our zeal to cut out all that is unreal we may not have used the knife too ruthlessly. Perhaps, indeed, reality is a child which cannot survive without its nurse illusion. But if so, that is of little concern to the scientist, who has good and sufficient reasons for pursuing his investigations in the world of shadows and is content to leave to the philosopher the determination of its exact status in regard to reality. In the world of physics we watch a shadowgraph performance of the drama of familiar life. The shadow of my elbow rests on the shadow table as the shadow ink flows over the shadow paper. It is all symbolic, and as a symbol the physicist leaves it. Then comes the alchemist Mind who transmutes the symbols. The sparsely spread nuclei of electric force become a tangible solid, their restless agitation becomes the warmth of summer; the octave of ætherial vibrations becomes a gorgeous rainbow. Nor does the alchemy stop here. In the transmuted world new significances arise which are scarcely to be traced in the world of symbols; so that it becomes a world of beauty and purpose – and, alas, suffering and evil.

ARTHUR EDDINGTON, *The Nature of the Physical World*, 1928
   (Gifford Lectures, 1927)

The problem of defining the relationship between mind and matter, and in particular between mind and body, has usually been thought of as difficult in the past because matter and the body have seemed so concrete, so substantial, and mind so insubstantial. Are we now to regard this quality of substantiality as a creation, a fiction – almost an attribute – of mind, and is it matter which turns out to be so abstract as to lead a doubtful existence?

11. Mind and Eye

Can then physics and chemistry out of themselves explain that a pin's-head ball of cells in the course of so many weeks becomes a child? They more than hint that they can. A highly competent observer, after watching a motion-film photo-record taken with the microscope of a cell-mass in the process of making bone, writes: 'Team-work by the cell-masses. Chalky spicules of bone-in-the-making shot across the screen, as if labourers were raising scaffold-poles. The scene suggested purposive behaviour by individual cells, and still more by colonies of cells arranged as tissues and organs.'[1] That impression of concerted endeavour comes, it is no exaggeration to say, with the force of a self-evident truth. The story of the making of the eye carries a like inference.

The eye's parts are familiar even apart from technical knowledge and have evident fitness for their special uses. The likeness to an optical camera is plain beyond seeking. If a craftsman sought to construct an optical camera, let us say for photography, he would turn for his materials to wood and metal and glass. He would not expect to have to provide the actual motor power adjusting the focal length or the size of the aperture admitting light. He would leave the motor power out. If told to relinquish wood and metal and glass and to use instead some albumen, salt, and water, he certainly would not proceed even to begin. Yet this is what that little pin's-head bud of multiplying cells, the starting embryo, proceeds to do. And in a number of weeks it will

[1] E. G. Drury, '*Psyche and the Physiologists*' and other Essays on Sensation (London, 1938), p. 4.

have all ready. I call it a bud, but it is a system separate from that of its parent, although feeding itself on juices from its mother. And the eye it is going to make will be made out of those juices. Its whole self is at its setting out not one ten-thousandth part the size of the eye-ball it sets about to produce. Indeed it will make two eye-balls built and finished to one standard so that the mind can read their two pictures together as one. The magic in those juices goes by the chemical names, protein, sugar, fat, salts, water. Of them 80 per cent is water . . .

The eye-ball is a little camera. Its smallness is part of its perfection. A spheroid camera. There are not many anatomical organs where exact shape counts for so much as with the eye. Light which will enter the eye will traverse a lens placed in the right position there. *Will* traverse; all this making of the eye which *will* see in the light is carried out in the dark. It is a preparing in darkness for use in light. The lens required is biconvex and to be shaped truly enough to focus its pencil of light at the particular distance of the sheet of photosensitive cells at the back, the retina. The biconvex lens is made of cells, like those of the skin but modified to be glass-clear. It is delicately slung with accurate centring across the path of light which *will* in due time some months later enter the eye. In front of it a circular screen controls, like the iris-stop of a camera or microscope, the width of the beam and is adjustable, so that in a poor light more is taken for the image. In microscope, or photographic camera, this adjustment is made by the observer working the instrument.[1] In the eye this adjustment is automatic, worked by the image itself!

The lens and screen cut the chamber of the eye into a front half and a back half, both filled with clear humour, practically water, kept under a certain pressure maintaining the eye ball's right shape. The front chamber is completed by a layer of skin specialized to be glass-clear, and free from blood-vessels which if present would with their blood throw shadows within the eye. This living glass-clear sheet is covered with a layer of tear-water constantly renewed. This tear-water has the special chemical

[1] *No longer invariably true. (Ed.)*

power of killing germs which might inflame the eye. This glass-clear bit of skin has only one of the fourfold set of skin-senses; its touch is always 'pain', for it should *not* be touched. The skin above and below this window grows into movable flaps, dry outside like ordinary skin, but moist inside so as to wipe the window clean every minute or so from any specks of dust, by painting over it fresh tear-water.

The light-sensitive screen at the back is the key-structure. It registers a continually changing picture. It receives, takes, and records a moving picture life-long without change of 'plate', through every waking day. It signals its shifting exposures to the brain.

This camera also focuses itself automatically, according to the distance of the picture interesting it. It makes its lens 'stronger' or 'weaker' as required. This camera also turns itself in the direction of the view required. It is moreover contrived as though with fore-thought of self-preservation. Should danger threaten, in a moment its skin shutters close, protecting its transparent window. And the whole structure, with its prescience and all its efficiency, is pro-duced by and out of specks of granular slime arranging themselves as of their own accord in sheets and layers, and acting seemingly on an agreed plan. That done, and their organ complete, they abide by what they have accomplished. They lapse into relative quietude and change no more. It all sounds an unskilful over-stated tale which challenges belief. But to faithful observation so it is. There is more yet.

The little hollow bladder of the embryo-brain, narrowing itself at two points so as to be triple, thrusts from its foremost chamber to either side a hollow bud. This bud pushes toward the overlying skin. That skin, as though it knew and sympathized, then dips down forming a cup-like hollow to meet the hollow brain-stalk growing outward. They meet. The round end of the hollow brain-bud dimples inward and becomes a cup. Concurrently, the ingrowth from the skin nips itself free from its original skin. It rounds itself into a hollow ball, lying in the mouth of the brain-cup. Of this stalked cup, the optic cup, the stalk becomes in a few weeks a cable of a million nerve fibres connecting the nerve-cells within the eyeball itself with the brain. The optic cup, at first just a two–deep layer of somewhat simple-looking cells, multiplies its layers at the bottom of the cup where, when light enters the eye – which will not be for some weeks yet – the photo-image will in due course lie. There the layer becomes a fourfold layer of great complexity. It is strictly speaking a piece of the brain lying within the eyeball . . .

The deepest cells at the bottom of the cup become a photo-sensitive layer – the sensitive film of the camera. If light is to act on the retina – and it is from the retina that light's visual effect is known to start – it must be absorbed there. In the retina a delicate purplish pigment absorbs incident light and is bleached by it, giving a light-picture. The photo–chemical effect generates nerve-currents running to the brain.

The nerve-lines connecting the photo-sensitive layer with the brain are not simple. They are in series of relays. It is the primitive cells of the optic cup, they and their progeny, which become in a few weeks these relays resembling a little brain, and each and all so shaped and connected as to transmit duly to the right points of the brain itself each light picture momentarily formed and 'taken'. On the sense-cell layer the 'image' has, picture-like, two dimensions. These space-relations 'reappear' in the mind; hence we may think their data in the picture are in some way preserved in the electrical patterning of the resultant disturbance in the brain. But reminding us that the step from electrical disturbance in the brain to the mental experience is the mystery it is, the mind adds the third dimension when interpreting the two-dimensional picture! Also it adds colour; in short it makes a three-dimensional visual scene out of an electrical disturbance.

All this the cells lining the primitive optic cup have, so to say, to bear in mind, when laying these lines down. They lay them down by becoming them themselves . . .

The human eye has about 137 million separate 'seeing' elements spread out in the sheet of the retina. The number of nerve-lines leading from them to the brain gradually condenses down to little over a million. Each of these has in the brain, we must think, to find its right nerve-exchanges. Those nerve-exchanges lie far apart, and are but stations on the way to further stations. The whole crust of the brain is one thick tangled jungle of exchanges and of branching lines going thither and coming thence. As the eye's cup develops into the nervous retina all this intricate orientation to locality is provided for by corresponding growth in the brain. To compass what is needed adjacent cells, although sister and sister, have to shape themselves quite differently the one from the other. Most become patterned filaments, set lengthways in the general direction of the current of travel. But some thrust out arms laterally as if to embrace together whole cables of the conducting system.

Nervous 'conduction' is transmission of nervous signals, in this case to the brain. There is also another nervous process, which physiology was slower to discover. Activity at this or that point in the conducting system, where relays are introduced, can be decreased even to suppression. This lessening is called inhibition; it occurs in the retina as elsewhere. All this is arranged for by the developing eye-cup when preparing and carrying out its million-fold connexions with the brain for the making of a seeing eye. Obviously there are almost illimitable opportunities for a false step. Such a false step need not count at the time because all that we have been considering is done months or weeks before the eye can be used. Time after time so perfectly is all performed that the infant eye is a good and fitting eye, and the mind soon is instructing itself and gathering knowledge through it. And the child's eye is not only an eye true to the human type, but an eye with personal likeness to its individual parent's. The many cells which made it have executed correctly a multitudinous dance engaging millions of performers in hundreds of sequences of particular different steps, differing for each performer according to his part. To picture the complexity and the precision beggars any imagery I have. But it may help us to think further.

There is too that other layer of those embryonic cells at the back of the eye. They act as the dead black lining of the camera; they with their black pigment kill any stray light which would blur the optical image. They can shift their pigment. In full daylight they screen, and at night they unscreen, as wanted, the special seeing elements which serve for seeing in dim light. These are the cells which manufacture the purple pigment, 'visual purple', which sensitizes the eye for seeing in low light.

Then there is that little ball of cells which migrated from the skin and thrust itself into the mouth of the eye-stalk from the brain. It makes a lens there; it changes into glass-clear fibres, grouped with geometric truth, locking together by toothed edges. The pencil of light let through must come to a point at the right distance for the length of the eye-ball which is to be. Not only must the lens be glass-clear but its shape must be optically right, and its substance must have the right optical refractive index. That index is higher than that of anything else which transmits light in the body. Its two curved surfaces back and front must be truly centred on one and the right axis, and each of the sub-spherical curvatures must be curved to the right degree, so that, the refractive index being right, light is brought to a focus on the retina and gives there a shaped image. The optician obtains glass of the desired refractive index and skilfully grinds its curvatures in accordance with the mathematical formulae required. With the lens of the eye, a batch of granular skin-cells are told off to travel from the skin to which they strictly belong, to settle down in the mouth of the optic cup, to arrange themselves in a compact and suitable ball, to turn into transparent fibres, to assume the right refractive index, and to make themselves into a subsphere with two correct curvatures truly centred on a certain axis. Thus it is they make a lens of the right size, set in the right place, that is, at the right distance behind the transparent window of the eye in front and the sensitive seeing screen of the retina behind. In short they behave as if fairly possessed.

. . . The cells composing the core of this living lens are denser than those at the edge. This corrects a focusing defect inherent in ordinary glass-lenses. Again, the lens of the eye, compassing what no glass-lens can, changes its curvature to focus near objects as well as distant when wanted, for instance, when we read. An elastic capsule is spun over it and is arranged to be cased by a special muscle. Further, the pupil – the camera stop – is self-adjusting. All this without our having even to wish it; without even our knowing anything about it, beyond that we are seeing satisfactorily.

The making of this eye out of self-actuated specks, which draw together and multiply and move as if obsessed with one desire, namely, to make the eye-ball. In a few weeks they have done so. Then, their madness over, they sit down and rest, satisfied to be safe-long what they have made themselves, and, so to say, wait for death.

The chief wonder of all we have not touched on yet. Wonder of wonders, though familiar even to boredom. So much with us that we forget it all our time. The eye sends, as we saw, into the cell-and-fibre forest of the brain throughout the waking day continual rhythmic streams of tiny, individually evanescent, electrical potentials. This throbbing streaming crowd of electrified shifting points in the spongework of the brain bears no obvious semblance in space-pattern, and even in temporal relation resembles but a little remotely the tiny two-dimensional upside-down picture of the outside world which the eye-ball paints on the beginnings of its nerve-fibres to the brain. But that little picture sets up an electrical storm. And that electrical storm so set up is one which effects a whole population of brain-cells. Electrical charges having in themselves not the faintest elements of the visual – having, for instance, nothing of 'distance', 'right-side-upness', nor 'vertical', nor 'horizontal', nor 'colour', nor 'brightness', nor 'shadow', nor roundness', nor 'squareness', nor 'contour', nor 'transparency', nor 'opacity', nor 'near', nor 'far', nor visual anything – yet conjure up all these. A shower of little electrical leaks conjures up for me, when I look, the landscape; the castle on the height, or, when I look at him, my friend's face, and how distant he is from me they tell me. Taking their word for it, I go forward and my other senses confirm that he is there.

It is a case of 'the world is too much with us'; too banal to wonder at. Those other things we paused over, the building and

shaping of the eye-ball, and the establishing of its nerve con-
nections with the right points of the brain, all those other things
and the rest pertaining to them we called in chemistry and physics
and final causes to explain to us. And they did so, with promise of
more help to come.

But this last, not the eye, but the 'seeing' by the brain behind
the eye? Physics and chemistry there are silent to our every
question. All they say to us is that the brain is theirs, that
without the brain which is theirs the seeing is not. But as to how?
They vouchsafe us not a word. Their negation goes further – they
assure us it is no concern of theirs at all.

CHARLES SHERRINGTON, *Man on his Nature*, 1940 (Gifford
Lectures, Edinburgh, 1937–8)

*Physics and chemistry may claim to be able out of themselves to
explain that a pin's-head ball of cells in the course of so many weeks
becomes a child. But we, and apparently Sherrington, cannot conceive of the process, still less describe it, without resorting to teleological
terminology and calling on 'final causes'. Small wonder that to
comprehend and discuss the process of 'seeing', let alone that of
'thinking', we have to fall back on terms such as 'mind' which have
no meaning in physics and chemistry.*

*Is this to deny the validity, the essential reality, of the processes
we are describing? Or is it severely to circumscribe the power of
physics and chemistry to describe these all-important processes – to
say that the 'real' description/explanation must be in terms other
than those of physics and chemistry? Or is it perhaps to acknowledge
that the human mind lacks the vocabulary, the concepts, the power
even, to grasp this 'real' explanation in anything other than partial
glimpses, which do not always seem to tally one with another?*

*It is hard to select a short extract from Sherrington (this one has
whole pages omitted from it as it is); his sentences and paragraphs
and chapters lead beguilingly on, never somehow quite reaching the
conclusion one is all the time expecting just round the next corner, but
passing so much of interest on the way.*

III. The Ghost in the Machine

The development of the individual mind therefore depends in part on the kind of brain which a man gets and in part on the kind of conscious experience through which it is his good fortune or bad fortune to pass. But it is very much 'in part'. We cannot explain individual mind development on these lines anything like completely, and there may well be other more important factors of which the scientific mind can as yet take no account and possibly never will be able to take account. Further, what is the nature of the mind – a different problem from that of its development – and what relationship does the mind bear to the living form of matter on which it is always based?

Science can tell us little yet. Psychology, clinical and experimental, accepts mind at 'face value' and, without inquiring into its nature, studies its workings. Electro-physiology describes the physico-chemical changes in the brain, but between them and their parallel at the mental level there is still a great gulf fixed. Para-psychology, the investigation of psychic phenomena, has only just started, and the results to date are not exciting. Some think that statistical studies have established a case for the occasional occurrence of telepathy (thought transference at a distance) with a reasonable degree of certainty, and some cases of precognition (foreknowledge of coming events) seem to have stood the test of scientific investigation. Further, some people seem to possess psychic powers. Although much is fraudulent, a hard core of para-psychological phenomena, stuff which cannot be explained in terms of any laws of nature known at present, does seem to remain. On the other hand, convincing evidence for the existence of mind or personality apart from the living body has not yet been obtained.

Science failing us, we are therefore forced into philosophy. Dualism in some form is always tempting. For at first sight it seems to fit into man's own self experience and is consistent with everyone's real conviction, however hard he may try to reason it away, that within certain limits he is free by an effort of will to control his own thoughts and actions. 'All human experience is in favour of free will,' said Dr. Johnson, 'all reason against it.' Dualism also fits in with the existence of absolute standards in art and morals. Real art reveals beauty divorced from the subjective judgement of the individual man or that common to his class or his generation. True moral sense reveals divine law rather than social necessity in the struggle for existence. Further, dualism is consistent with the concept of some form of personal survival after death and in keeping with the teaching of revealed religion. But dualism in its crude form – body and mind as two separate things – does not fit the facts of physiology or the psycho-somatic concept of the development of the individual man as outlined in this chapter. How can two separate things be so intimately related?

One way out of this dilemma is materialism, that is to say the point of view which maintains that consciousness and mind are the result of the behaviour of matter. According to this view, not to be confused with agnosticism (just not knowing), the mind must cease with the death of the body. All human conduct is purely reflex and predictable, rendering the concept of free will meaningless, and reducing all human behaviour to the dead level of determinism. But materialism cuts little ice today. For physical research has revealed that modern matter, instead of being the good sound solid stuff which the scientists of the last century confidently supposed, consists of protons, neutrons, and electrons as far apart among themselves as the planets and the sun, and containing a vast store of hitherto completely unsuspected energy locked up inside it. Further, it is quite clear that material things do not possess sensory qualities, colour, taste, smell, etc., in 'their own right'. Rather these are added to the otherwise drab purely physical world when it collides with the sensory machine, also made of protons and electrons, of the human nervous system. Alter his sensory machine and the world is different to the man. The colour-blind see it drab, and to the deaf the noisy city is as silent as the grave. Modern matter has become as inexplicable as the mind which it was intended to explain.

The other alternative is to deny matter and attempt to explain all human experience in terms of mind. 'I think, therefore I am' was the starting point of Descartes' system of philosophy, and a hundred years later Bishop Berkeley pushed this way of beginning to its logical conclusion. The only thing of which a man is absolutely certain, he maintained, is his own mind. All else is indirect inference from experience, and even today extreme idealism, as thus stated, is difficult to controvert. Nevertheless, the tide of common sense flows hard against it. 'I refute him thus,' exclaimed Dr. Johnson, kicking his foot against a stone. The plain man will not readily reject the evidence of his senses. A real world must exist, although it seems unlikely that the human mind can comprehend its true nature.

So as neither dualism, materialism, nor idealism will do, there is only one way out left, namely welding all the four main aspects of human experience – matter, energy, life, and mind – into one single universal whole. Shall we ever do it? That we possess the necessary cerebral machinery for the process sometimes seems improbable. (The idea that the human mind must necessarily be able to understand the real nature of things is a considerable assumption.) Nevertheless, a beginning has perhaps been made. Energy and matter, until recently regarded as separate and distinct, are now known to be capable of conversion into one another. Further, physicists can calculate by mathematics (which the human mind has invented) how much energy can be got out of so much matter and how much matter out of so much energy. And life and mind? Where do they stand in relation to each other and the other two? We understand neither life nor mind nor their relation to energy and matter as yet. So in our ignorance there is room for faith, whatever that is and wherever it comes from, in those who have it, and ground for hope for all, which in health at least, 'springs eternal in the human breast'.

A. E. CLARK-KENNEDY, *Human Disease*, 1957

Perhaps we need Planck to show us how to 'evade the hypnosis of the insoluble problem'.

IV. A Phantom Problem

It is . . . an experience annoying as can be, to find, after a long time spent in toil and effort, that the problem which has been preying on one's mind is totally incapable of solution at all – either because there exists no indisputable method to unravel it, or because considered in the cold light of reason, it turns out to be absolutely void of all meaning – in other words it is a *phantom problem*, and all that mental work and effort was expended on a mere nothing. There are many such phantom problems – in my opinion, far more than one would ordinarily suspect – even in the realm of science . . .

My first example is a phantom problem, for the triviality of which I beg your forgiveness. The room in which we now sit has two side walls, a right-hand one and a left-hand one. To you, *this* is the right side, to me, sitting facing you, *that* is the right side. The problem is: Which side is in reality the right-hand one? I know that this question sounds ridiculous, yet I dare call it typical of an entire host of problems which have been, and in part still are, the subject of earnest and learned debates, except that the situation is not always quite so clear . . .

Now let us proceed to the consideration of a problem which has always been regarded as of central importance because of its meaning to human life – the famous body-mind problem . . . What is the nature of the interrelation of physical and mental processes? Are mental processes caused by physical ones? And if so, according to what laws? How can something material act on something immaterial, and *vice versa*?

. . . In order to get to the bottom of the matter, let us ask ourselves this question of basic significance: Just what do we know about mental processes? In what circumstances and in what sense may we speak of mental processes? Let us consider first where we come across mental processes in this world. We must take it for granted that members of the higher animal kingdom as well as human beings have emotions and sensations. But as we descend to the lower animals – where is the border-line where sensation ceases to exist? Has a worm any sensation of pain as it is crushed under our feet? And may plants be considered capable of some kind of sensation? There are botanists who are disposed to answer this question affirmatively. But such a theory can never be put to the test, let alone proved, and the wisest course seems to be not to venture any opinion in this regard. Along the entire ladder of evolution, from the lowest order of life up to Man, there is no point at which one can establish a discontinuity in the nature of mental processes.

It is nevertheless possible to specify a quite definite borderline, of decisive importance for all that follows. This is the borderline between the mental processes within other individuals and the mental processes within one's own Ego. For everybody experiences his own emotions and sensations directly. They just simply exist for him. But we do not experience directly the sensations of any other individuals, however certain their existence may be, and we can only infer them in analogy to our own sensations. To be sure, there are physicians who solemnly claim to be able to perceive the emotions and moods of their patients no less clearly than the latter themselves. But such a claim can never be proved indisputably. Its questionability becomes most striking if we think of certain specific instances. Even the most sensitive dentist cannot feel the piercing pains which his patient at times has to suffer under his treatment. He can ascertain them only indirectly, on the basis of the moans or squirming of the patient. Or, to speak of a more pleasant situation, such as for instance a banquet, however clearly one may sense the pleasure of one's neighbor over the taste of his favorite wine, it is something quite different from tasting it on one's own tongue. What *you* feel, think, want, only *you* can know as first hand information. Other people can conclude it only indirectly, from your words, conduct, actions and mannerisms. When such physical manifestations are entirely absent, they have no basis whatever to enable them to know your momentary mental state.

This contrast between first-hand, or direct, and second-hand, or indirect, experience is a fundamental one. Since our primary aim is to gain direct, first-hand experience, we shall now discuss the interrelation of our mental and physical states.

First of all, we find that we may speak of conscious states only. To be sure, many processes, perhaps even the most decisive ones, must be taking place in the subconscious mind. But these are beyond the reach of scientific analysis. For there exists no science of the unconscious, or subconscious, mind. It would be a contradiction in terms, a self-contradiction. One does not know that which is subconscious. Therefore, all problems concerning the subconscious are phantom problems.

Let us therefore take a simple conscious process involving body and mind. I prick my hand with a needle, and feel a sensation of pain. The wound made by the pin is the physical element, the sensation of pain is the mental element of the process. The wound is seen, the pain is felt. Is there, then, an indisputable method of throwing light upon the interrelation of the two elements of this process? It is easy to realize that this is absolutely impossible. For there is nothing here upon which light is to be thrown. The visual perception of the wound and the feeling of pain are elementary facts of experience, but they are as different in nature as knowledge and feeling. Therefore, the question as to their essential interrelation represents no meaningful problem – it is just a phantom problem.

It is obvious that the two occurrences, the pin-prick and the sensation of pain, can be examined and analyzed most thoroughly, in every detail. But such an analysis calls for two different methods, which mutually preclude each other. Each of the two corresponds to one of two different viewpoints. In the following I will refer to them, respectively, as the *psychological* and the *physiological* viewpoints. Observation based on the psychological viewpoint is rooted in self-consciousness; therefore, it is applicable directly only to the analysis of one's own mental processes. On the other hand, observation based on the physiological viewpoint is directed at the processes in the external world; therefore, its direct scope is limited to physical processes. These two viewpoints are incompatible. The adoption of one when the other is called for, always leads to confusion. We cannot judge our mental processes directly from the physiological viewpoint any more than we can examine a physical process from the psychological viewpoint. This state of affairs makes the body-mind problem appear in a different light. For the examination of psychosomatic processes will yield entirely different results, according as the psychological or the physiological viewpoint is taken as the basis of observation. The psychological viewpoint will permit us to gain first-hand knowledge solely and exclusively of something that relates to our mental processes. The physiological viewpoint will produce first-hand information about physical processes only. It is therefore impossible to gain first-hand information about both physical and mental processes from any single viewpoint; and since in order to reach a clear conclusion, we must adhere to a given viewpoint, which automatically excludes the other, the search for the interrelation of physical and mental processes loses its meaning. In this case, there exist only physical processes *or* mental processes, but never processes which are physical *and* mental.

Therefore, it will do no harm to say that the physical and the mental are in no way different from each other. They are the selfsame processes, only viewed from two diametrically opposite directions. This statement is the answer to the riddle . . . , namely, how is one to conceive the fact that two types of processes so different from each other as the physical and the mental are so closely interlinked. The link has now been disclosed. At the same time, the body-mind problem has been recognised as another phantom problem.

MAX PLANCK, *Phantom Problems in Science* (In *Scientific Autobiography and Other Papers*), 1950 (tr. Frank Gaynor; first published in German 1947)

CHAPTER FIVE
The Nature of Science

I. Ants, Spiders and Bees

Those who have handled sciences have been either men of experiment or men of dogmas. The men of experiment are like the ant; they only collect and use; the reasoners resemble spiders, who make cobwebs out of their own substance. But the bee takes a middle course; it gathers its material from the flowers of the garden and of the field, but transforms and digests it by a power of its own. . . . Therefore from a closer and purer league between these two faculties, the experimental and the rational (such as has never yet been made), much may be hoped.

FRANCIS BACON, *Novum Organum*, 1620 (tr. R. Ellis and James Spedding 1857)

Of the two approaches to science he here describes, Bacon is best known for his emphasis on the experimental, factual one. And indeed the distinguishing mark of science since the time of Galileo and Bacon is usually taken to be its rigorous insistence on experimental verification.

The next passage shows a famous experimenter at work.

II. Experiments

There was taken a Wier, which being bent almost in the form of a Screw, constituted such an Instrument, to contein Coals and leave them every way accessible to the Air, as the tenth Figure declareth; the breadth of this Vessel was no less then that it might with ease be convey'd into the Receiver;[1] And having fill'd it to the height of about five Inches with throughly kindled Wood-coals, we[2] let it down into the Glass; and speedily closing it, we caus'd the

---

[1] *A glass-walled container with a pump for extracting the air. (Ed.)*
[2] *Boyle's assistant in these experiments was Robert Hooke. (Ed.)*

Pumper to ply his work, and observ'd that upon the very first exsuction of the Air (though perhaps not because of that onely) the Fire in the Coals began to grow very dim, and though the agitation of the Vessel did make them swing up and down (which in the free Air would have retarded the extinction of the Fire) yet when we could no longer discern any rednesse at all in any of them, casting our eyes upon a minute Watch we kept by us on this occasion, we found that from the beginning of the Pumping (which might be about two minutes after the Coals had been put in glowing) to the totall dis-appearing of the Fire, there had passed but three minutes.

Whereupon, to try the Experiment a little farther, we presently took out the Coals, in which it seems there had remain'd some little parcels of Fire, rather cover'd then totally quench'd: For in the open Air the Coals began to be re-kindled in several places, wherefore having by swinging them about in the Wier, throughly lighted them the second time, we let them down again into the Receiver, and clos'd it speedily as before; and then waiting till the Fire seem'd totally extinct without meddling with the Pump, we found that from the time the Vessel was clos'd, till that no fire at all could be perceiv'd, there had passed about four minutes: Whereby it seem'd to appear that the drawing away of the Ambient Air made the Fire to go out sooner then otherwise it would have done; though that part of the Air that we drew out left the more room for the stifling steams of the Coals to be receiv'd into.[1]

Lastly, Having taken out the Wier and put other coals into it, we did, in the same Room where the Engine stood, let it hang quietly by a string in the open Air to try how long the Fire would last without agitation, when no Air was kept from it. And we found that the Fire began to go out first at the top and outsides of the Coals; but inwards and near the bottom the Fire continu'd visible for above half an hour, a great part of the Coals, especially

[1] *In a later experiment, Boyle, having tried placing a smouldering wick in his Receiver and removing the air, repeats the process, but includes a sealed bladder of air. This expands as the pressure decreases, so that there is not, in fact, 'more room for the stifling steams' but merely less 'Ambient Air'. (Ed.)*

those next the bottom, being burnt to ashes before the Fire went out.

We caus'd likewise a piece of Iron to be forg'd of the bignesse of a midle siz'd Char-coal, and having made it red-hot throughout, we caus'd it in the lately mention'd Wier, to be speedily convey'd and shut up into the Receiver, being desirous to try what would become of a glowing Body, by reason of its texture more vehemently hot then a burning Coal of the same bignesse, and yet unlikely to send forth such copious and stifling Fumes. But we could not observe any manifest change upon the exsuction of the Air. The Iron began indeed to lose its Fiery redness at the top, but that seem'd to be because it was at the upper end somewhat more slender then at the lower: The redness, though it were in the day time, continu'd visible about four minutes; and then, before it did quite dis-appear, we turn'd the Key of the Stop-cock, but could not discern any change of the Iron upon the rushing in of the Air.

ROBERT BOYLE, *New Experiments Physico-Mechanical Touching the Spring of the Air and its Effects*, 1660

Boyle has here, to borrow a phrase from the writer of the next passage, no 'conceptual scheme' into which to fit his results, or at best a wholly irrelevant one. His experiments are painstaking but random, undirected by a suitable theory; in other words, they lack a pre-conceived hypothesis to test and either establish or overthrow. Additions to human knowledge will in such circumstances be almost entirely negative.

In fact, not for another century was the problem of combustion to be solved. Boyle helped finally to discredit the idea of an element of fire, but it had a more sophisticated successor in phlogiston, a mythical substance which was supposed to escape from a body when it burned and be absorbed by the air. Air which would no longer support combustion, or only with difficulty, was said therefore to be saturated with phlogiston, while Priestly, the discoverer of oxygen, persisted in calling the gas 'dephlogisticated air' because it was able to absorb phlogiston so much more readily than ordinary air. And all this in the teeth of experimental evidence which showed that a body gained rather than lost in weight when it gave up its phlogiston. Indeed, some apologists went so far as to attribute a negative weight to phlogiston.

The next passage deals with the relationship between fact, or experimental evidence, and theory, or 'conceptual schemes'.

III. Planned Experiments

Galileo related in his *Dialogues Concerning Two New Sciences*, published in 1638, that a lift pump which failed to work was once called to his attention. The machine was in perfect order but it had been asked to do what a workman familiar with pumps said was an impossible task, namely, draw up water from a cistern for a distance of more than approximately 34 feet. Probably practical men long before this day must have recognized that there were limits to what a suction or lift pump could do. For example, in one of the illustrations in the famous work by Agricola on Mining published in 1556 several lift pumps are shown one above the other; this arrangement was required because of the limitation which was called to Galileo's attention.

Now clearly the Aristotelian concept of nature's abhorring a vacuum required amendment as to why this abhorrence only extended to a limited distance. But Galileo, though recognizing that a general principle was involved, missed the next step. He believed that water would not rise about 34 feet in a lift pump because the column broke of its own weight. He had the wrong analogy in mind – he was comparing the behaviour of a column of water to the breaking point of a copper wire. Here is an example of careful observation, in a sense experimentation (assuming that Galileo tried various pumps at various heights above the water level), and yet a fruitless concept issues even from one of the very great scientists of all times. . . .

Two of Galileo's disciples, Torricelli and Viviani, did find the fruitful trail. But before recounting their story, let us remember that the conceptual scheme implied by the phrase 'nature abhors a vacuum' was by no means the nonsense we sometimes imply today. In a limited way this idea explained adequately a number of apparently unrelated phenomena and that is one of the tests of any conceptual scheme. For example, it explained the action of the lift pumps in common use, the adhesion of one piece of wet marble to another, the action of a bellows, one's inability to make a 'hole' in a liquid the way one can in a solid, and so on. The fact that water cannot be raised by suction more than thirty-four feet seemed to require an extension of the then current concept, not necessarily a revolutionary new concept. Usually our conceptual schemes grow by an evolutionary process, by the gradual incorporation of a series of amendments, so to speak. In this case a completely new idea came along and rendered obsolete the older one. We shall meet the same phenomenon later in the next chapter when we trace the overthrow of the phlogiston theory.

We can put it down as one of the principles learned from the history of science that a theory is only overthrown by a better theory, never merely by contradictory facts. Attempts are first made to reconcile the contradictory facts to the existing conceptual scheme by some modification of the concept. Only the combination of a new concept with facts contradictory to the old ideas finally brings about a scientific revolution. And when once this has taken place, then in a few short years discovery follows upon discovery and the branch of science in question progresses by leaps and bounds.

With this principle in mind, let us turn back to Italy in the fifth decade of the seventeenth century. Galileo and his pupils in Florence are turning over in their minds various explanations of why nature's abhorrence of a vacuum had a rather sharp and definite limit. At some time and in some way not recorded, one of them, Torricelli, connected this fact with another fact which was generally accepted even at that time, namely, that air had weight. If air had weight, why might it not exert pressure on the surface of the water in a well and thus force the water up the lift pump as the piston rises and produces suction? The height of 34 feet of water would thus represent the weight of water which this pressure of the air on the surface of the earth could maintain. Apparently reasoning thus, Torricelli, a young man of 35 in 1643, and Viviani, aged 21, tried some experiments with another liquid, namely, mercury, which is about 14 times as heavy as water. If the idea they were developing were correct, the pressure of air which enveloped the earth would hold up a column of mercury only about 1/14 as high as that of water, namely, a little more than 2 feet tall. A column of this height was something that could be managed with ease. They took a glass tube about a finger's width in diameter and about three feet long, sealed it off at one end, filled it completely with mercury, and keeping a finger over the end inverted it in an open vessel of mercury.

The expected happened, as it sometimes does in this type of experimentation when men of ability plan it in advance. The mercury fell to a height of approximately 30 inches above the level of the mercury in the open vessel or trough. The space above in the tube appeared empty. For the first time the world had seen a vacuum, to speak loosely; or, to express the result of the experiment in more accurate terms, a vacuum had been created in the upper end of the tube because the pressure of the atmosphere could support a column of mercury only about 30 inches long. Such a vacuum became known at once as the Torricellian vacuum.

JAMES B. CONANT, *On Understanding Science*, 1947

Armed with the conceptual scheme whose discovery Conant has just described, Boyle was able to postulate an hypothetical extension to or refinement of the said conceptual scheme, and then carry out an experiment to test his theory – an experiment which was to establish the law to which Boyle gave his name.

IV. The Law

We took then a long Glass-Tube, which by a dexterous hand and the help of Lamp was in such a manner crooked at the bottom, that the part turned up was almost parallel to the rest of the Tube, and then the Orifice of this shorter leg of the Siphon (if I may so call the whole Instrument) being Hermetically seal'd, the length of it was divided into Inches, (each of which was subdivided into eight parts) by a straight list of paper, which containing those Divisions was carefully pasted all along it: then putting in as much Quicksilver as served to fill the Arch or bended part of the Siphon, that the *Mercury* standing in a level might reach in the one leg to the bottom of the divided paper, and just to the same height or Horizontal line in the other; we took care, by frequently inclining the Tube, so that the Air might freely pass from one leg into the other by the sides of the *Mercury*, (we took (I say) care) that the Air at last included in the shorter Cylinder should be of the same laxity with the rest of the air about it. This done we began to pour Quicksilver into the longer leg of the Siphon, which by its weight pressing up that in the shorter leg, did by degrees streighten the included Air: and continuing this pouring of Quicksilver till the Air in the shorter leg was by condensation reduced to take up but half the space it possess'd (I say, *possess'd*, not *fill'd*) before; we cast our eyes upon the longer leg of the Glass, on which was likewise pasted a list of Paper carefully divided into Inches and parts, and we observed, not without delight and satisfaction, that the Quicksilver in that longer part of the Tube was 29. Inches higher then the other. Now that this Observation does both very well agree with and confirm our *Hypothesis*, will be easily discerned by him that takes notice that we teach, and Monsieur Paschall and our English friends Experiments prove,[1] that the greater the weight is that leans upon the air, the more forcible is its endeavour of Dilation, and consequently its power of resistance, (as other Springs are stronger when bent by greater weights.) For this being considered, it wil appear to agree rarely-well with the Hypothesis, that as according to it the Air in that degree of density and correspondent measure of resistence to which the weight of the incumbent Atmosphere had brought it, was able to counterbalance and resist the pressure of a Mercurial Cylinder of about 29. Inches, as we are taught by the *Torricellian* Experiment;

[1] *Robert Hooke and Lord Brouncker.* (*Ed.*)

so here the same Air being brought to a degree of density about twice as great as that it had before, obtains a Spring twice as strong as formerly. As may appear by its being able to sustain or resist a Cylinder of 29. Inches in the longer Tube, together with the weight of the Atmospherical Cylinder, that lean'd upon those 29. Inches of Mercury; and, as we just now inferr'd from the Torricellian Experiment, was equivalent to them.

ROBERT BOYLE, *A Defence of the Doctrine touching the Spring and Weight of the Air*, 1662

Faraday, too, had a conceptual scheme which he was able to narrow down to an hypothesis specific enough to be put to the test. Moreover, he was able to design and carry out the relatively sophisticated experiment this required, and lastly to interpret its results with scrupulous honesty.

v. Gravity and Electricity

2702 The long and constant persuasion that all the forces of nature are mutually dependent, having one common origin, or rather being different manifestations of one fundamental power, has made me often think upon the possibility of establishing, by experiment, a connexion between gravity and electricity, and so introducing the former into the group, the chain of which, including also magnetism, chemical force and heat, binds so many and such varied exhibitions of force together by common relations. . . .

2703 In searching for some principle on which an experimental inquiry after the identification or relation of the two forces could be founded, it seemed that if such a relation existed, there must be something in gravity which would correspond to the dual or antithetical nature of the forms of force in electricity and magnetism. . . .

2704 The thought on which the experiments were founded was, that, as two bodies moved towards each other by the force of gravity, currents of electricity might be developed either in them or in the surrounding matter in one direction; and that as they were by extra force moved from each other against the power of gravitation, the opposite currents might be produced. . . . As the effect looked for was exceedingly small, so no hope was entertained of any result except by means of the gravitation of the earth. The earth was therefore made to be the one body, and the indicating mass of matter to be experimented with the other.

2705 First of all, a body, which was to be allowed to fall, was surrounded by a helix, and then its effect in falling sought for. Now a body may either fall with a helix or through a helix. Covered copper wire, to the amount of 350 feet in length, was made into a hollow cylindrical helix, about 4 inches long, its internal diameter being 1 inch and its external diameter 2 inches. It was attached to a line running upon an easy pulley, so that it could be raised 36 feet, and then allowed to fall with an accelerated velocity on to a very soft cushion, its axis remaining vertical the whole time. Long covered wires were made fast to its two extremities, and these being twisted round each other, were attached to a very delicate galvanometer, placed about 50 feet aside from the line of fall, and on a level midway with its course. The accuracy of the connexion and the direction of the set of the needle, were then both ascertained by the introduction of a feeble thermo–electric combination into the current.[1] Such a helix, either in rising or falling, can produce no deviation at the galvanometer by any current due to the magnetism of the earth; for as it remains parallel to itself during the fall, so the lines of equal magnetic force, which being parallel to the dip, are intersected by the wire convolutions of the descending helix, are cut with an equal velocity on both sides of the helix, and consequently no effect of magneto–electric induction is produced. Neither in rising nor in falling did this helix present any trace of action at the galvanometer; whether the connexion with the galvanometer was continued the whole time, or whether it was cut off just before the diminution or cessation of motion either way, or whether the

---

[1] *Circuit.* (*Ed.*)

rising and falling were made to occur isochronously with the times of vibration of the galvanometer needle. So, though no effect of gravity appeared in the helix itself, still no source of error appeared to arise in this mode of using it.

2706 A solid cylinder of copper, three-fourths of an inch in diameter and 7 inches in length, was now introduced into the helix and carefully fastened in it, being bound round with a cloth so as not to move, and this compound arrangement was allowed to fall as before (2705). It gave very minute but remarkably regular indications of a current at the galvanometer; and the probability of these being related to gravity appeared the greater, when it was found that on raising the helix or core, similar indications of contrary currents appeared. It was some time before I was able to refer these currents to their true cause, but at last I traced them to the action of a part of the connecting wires proceeding from the helix to the galvanometer. The two wires had been regularly twisted together, but the effect of many falls had opened a part near the middle distance into a sort of loop, so that the wires, instead of being tightly twisted together like the strands of a rope, were separate for 3 feet, as if the strands were open. In falling, this loop opened out more or less, but always in the same manner; and the consequence was that the part of it representing the transverse opening, which was furthest from the galvanometer, travelled over a larger space than the corresponding part nearest the galvanometer. Now had they travelled through equal spaces, the effect of the magnetic lines of force of the earth upon them would have been equal, and no effect at the galvanometer would have been produced; as it was, currents in opposite directions, but of unequal amounts of force, tended to be produced, and a current equal to the difference actually appeared. Such a case is described in my earliest researches on terrestrial magno-electro induction (171). It is evident that the current should appear in the reverse direction, as the helix and wires are raised in the air, and thus arose the reverse effect described above. Therefore no positive or favourable evidence was supplied in favour of the original assumption by this use of a copper core in the helix.

2707 The copper was selected as a heavy body and an excellent conductor of electricity. On its dismissal, a bismuth cylinder of equal size was employed to replace it as a substance eminently diamagnetic, and a bad conductor amongst metals. Uncertain evidence arose; but by close attention, first to one point and then to another, all the indications disappeared, and then the rising or falling of the bismuth produced no effect on the galvanometer.

2708 An *iron* cylinder was also employed as a magnetic metal, but when made perfectly secure, so as to prevent any motion relative to the helix, it was equally indifferent with the copper and bismuth.

2709 Cylinders of glass and shell-lac were employed as non-conducting substances, but without effect.

2710 In other experiments the helix was *fixed*, and the different substances in the form of cylinders, three-fourths of an inch in diameter and 24 inches long, were dropped through it, or else raised through it with an accelerated velocity; but in neither case was any effect produced. Rods of copper, bismuth, glass, shell-lac and sulpher were employed. Occasionally these rods were made to rotate rapidly before and during their fall; and many other conditions were devised and carried into effect, but always with negative results, when sources of error were avoided or accounted for.

MICHAEL FARADAY, *Experimental Researches in Electricity, Vol. III*, 1855

A theory in a science such as astronomy, geology or that branch of biology that deals with the development of life in past ages is open to verification by observation, but not usually to verification by experiment. In other words, theories must still be shown to fit the facts, but the facts are here ascertained by studying the results of those experiments which nature herself has been and often still is carrying out. (Indeed, it is arguable that in the case of small islands, and in particular the Galapagos Islands, nature has gone so far as to set up laboratories in which the workings of evolution can be studied under conditions strangely analogous to those of a scientific experiment.)

There follow two accounts of how the same hypothesis – that natural selection is the means, the mechanism, whereby evolutionary developments are effected–came to be formulated quite independently by Charles Darwin and A. R. Wallace. This was preceded, particularly in the case of Darwin, by the amassing of a great many facts, in which task he was guided by the 'conceptual scheme' or theory of evolution (which had been in the air to Darwin's knowledge since the time of Lamarck and his own grandfather, Erasmus Darwin, and in fact for some years before that). What is most striking in these two extracts, perhaps, is how a climate of opinion favourable to the idea of evolution by natural selection had been growing up, how ideas in the work of other scientists and thinkers led towards the same conclusion, and how two independent minds reached that conclusion by similar but not identical routes.

VI. The Struggle for Existence

From September 1854 onwards I devoted all my time to arranging my huge pile of notes, to observing, and experimenting, in relation to the transmutation of species. During the voyage of the *Beagle* I had been deeply impressed by discovering in the Pampean formation great fossil animals covered with armour like that on the existing armadillos; secondly, by the manner in which closely allied animals replace one another in proceeding southwards over the Continent; and thirdly, by the South American character of most of the productions of the Galapagos archipelago, and more especially by the manner in which they differ slightly on each island of the group; none of these islands appearing to be very ancient in a geological sense.

It was evident that such facts as these, as well as many others, could be explained on the supposition that species gradually become modified; and the subject haunted me. But it was equally evident that neither the action of the surrounding conditions, nor the will of the organisms[1] (especially in the case of plants), could account for the innumerable cases in which organisms of every

[1] *A rather unfair reference to Lamarck's views.* (*Ed.*)

kind are beautifully adapted to their habits of life, – for instance, a woodpecker or tree-frog to climb trees, or a seed for dispersal by hooks or plumes. I had always been much struck by such adaptations, and until these could be explained it seemed to me almost useless to endeavour to prove by indirect evidence that species have been modified.

After my return to England it appeared to me that by following the example of Lyell in Geology, and by collecting all facts which bore in any way on the variation of animals and plants under domestication and nature, some light might perhaps be thrown on the whole subject. My first note-book was opened in July 1837. I worked on true Baconian principles, and without any theory collected facts on a wholesale scale, more especially with respect to domestic productions, by printed enquiries, by conversation with skilful breeders and gardeners, and by extensive reading. When I see the list of books of all kinds which I read and abstracted, including whole series of Journals and Transactions, I am surprised at my industry. I soon perceived that selection was the keystone of man's success in making useful races of animals and plants. But how selection could be applied to organisms living in a state of nature remained for some time a mystery to me.

In October 1838, that is, fifteen months after I had begun my systematic enquiry, I happened to read for amusement Malthus on *Population*, and being well prepared to appreciate the struggle for existence which everywhere goes on from long-continued observation of the habits of animals and plants, it at once struck me that under these circumstances favourable variations would tend to be preserved, and unfavourable ones to be destroyed. The result of this would be the formation of new species. Here, then, I had at last got a theory by which to work; but I was so anxious to avoid prejudice, that I determined not for some time to write even the briefest sketch of it. In June 1842 I first allowed myself the satisfaction of writing a very brief abstract of my theory in pencil in 35 pages; and this was enlarged during the summer of 1844 into one of 230 pages, which I had fairly copied out and still possess.

But at that time I overlooked one problem of great importance; and it is astonishing to me, except on the principle of Colombus and his egg,[1] how I could have overlooked it and its solution. This problem is the tendency in organic living beings descended from the same stock to diverge in character as they become modified. That they have diverged greatly is obvious from the manner in which species of all kinds can be classed under genera, genera under families, families under sub-orders, and so forth; and I can remember the very spot in the road, whilst in my carriage, when to my joy the solution occurred to me; and this was long after I had come to Down.[2] The solution, as I believe, is that the modified offspring of all dominant and increasing forms tend to become adapted to many and highly diversified places in the economy of nature.

Early in 1856 Lyell advised me to write out my views pretty fully, and I began at once to do so on a scale three or four times as extensive as that which was afterwards followed in my 'Origin of Species'; yet it was only an abstract of the materials which I had collected, and I got through about half the work on this scale. But my plans were overthrown, for early in the summer of 1858 Mr Wallace, who was then in the Malay archipelago, sent me an essay 'On the Tendency of Varieties to depart indefinitely from the Original Type'; and this essay contained exactly the same theory as mine. Mr Wallace expressed the wish that if I thought well of his essay, I should send it to Lyell for perusal.

The circumstances under which I consented at the request of Lyell and Hooker to allow of an abstract from my MS., together with a letter to Asa Gray, dated September 5, 1857, to be published at the same time with Wallace's Essay, are given in the 'Journal of the Proceedings of the Linnean Society', 1858, p. 43. I was at

---

[1] *Asked whether someone else would not have discovered America had he not done so, Colombus is said to have invited his hearers to stand an egg upright on its its end. When all had failed, he took the egg himself and brought it down on the table hard enough to crush the shell slightly at one end, then left it standing there. (Ed.)*

[2] *Where Darwin had moved to arrange his 'huge pile of notes'. (Ed.)*

first very unwilling to consent, as I thought Mr. Wallace might consider my doing so unjustifiable, for I did not then know how generous and noble was his disposition. The extract from my MS. and the letter to Asa Gray had neither been intended for publication and were badly written. Mr Wallace's essay, on the other hand, was admirably expressed and quite clear. Nevertheless, our joint productions excited very little attention, and the only published notice of them which I can remember was by Professor Haughton of Dublin, whose verdict was that all that was new in them was false, and what was true was old. This shows how necessary it is that any new view should be explained at considerable length in order to arouse public attention.

CHARLES DARWIN, *Autobiography* (Chapter II, Vol. 1, *Life and Letters of Charles Darwin*, ed. Francis Darwin, 1888)

VII. The Survival of the Fittest

While thinking (as I had thought for years) over the possible causes of the change of species, the action of these 'positive checks' to increase, as Malthus termed them, suddenly occurred to me. I then saw that war, plunder and massacres among men were represented by the attacks of carnivora on herbivora, and of the stronger upon the weaker among animals. Famine, droughts, floods and winter's storms, would have an even greater effect on animals than on men; while as the former possessed powers of increase from twice to a thousand-fold greater than the latter, the ever-present annual destruction must also be many times greater.

Then there flashed upon me, as it had done twenty years before upon Darwin, the *certainty* that those which, year by year, survive this terrible destruction must be, on the whole, those which had some little superiority enabling them to escape each special form of death to which the great majority succumbed – that, in the well-known formula, the fittest would survive. Then I at once saw, that the ever present *variability* of all living things would furnish the material from which, by the mere weeding out of those less adapted to the actual conditions, the fittest alone would continue the race. But this would only tend to the persistence of those best adapted to the actual conditions; and on the old idea of the permanence and practical unchangeability of the inorganic world, except for a few local and unimportant catastrophes, there would be no necessary change of species.

But along with Malthus I had read, and been even more deeply impressed by, Sir Charles Lyell's immortal 'Principles of Geology', which had taught me that the inorganic world – the whole surface of the earth, its seas and lands, its mountains and valleys, its rivers and lakes, and every detail of its climatic conditions, were and always had been in a continual state of slow modification. Hence it became obvious that the forms of life must have become continually adjusted to these changed conditions in order to survive. The succession of fossil remains throughout the whole geological series of rocks is the record of this change; and it became easy to see that the extreme slowness of these changes was such as to allow ample opportunity for the continuous automatic adjustment of the organic to the inorganic world, as well as of each organism to every other organism in the same area, by the simple process of 'variation and survival of the fittest'.

ALFRED RUSSEL WALLACE, *The Darwin-Wallace Celebration held on 1 July 1908 by the Linnean Society of London*

Finally, a return in Whitehead's words to what Bacon said at the outset. Note how language, too, depends on a traffic and a tension between 'general principles' and 'irreducible and stubborn facts': witness Bacon's use of simile, and Whitehead's of words like tinge, vehement, stubborn, absorbed, weaving, sporadically, balance, infects, cultivated, salt *and* sweet.

VIII. Facts and Theories

This new tinge to modern minds is a vehement and passionate interest in the relation of general principles to irreducible and stubborn facts. All the world over and at all times there have been practical men, absorbed in 'irreducible and stubborn facts'; all the world over and at all times there have been men of philosophic temperament who have been absorbed in the weaving of general principles. It is this union of passionate interest in the detailed facts with equal devotion to abstract generalisation which forms the novelty in our present society. Previously it had appeared sporadically and as if by chance. This balance of mind has now become part of the tradition which infects cultivated thought. It is the salt which keeps life sweet.

A. N. WHITEHEAD, *Science and the Modern World*, 1926

CHAPTER SIX
Likenesses

I. The Solar System

Because of the way it came into existence,[1] the solar system has only one-way traffic – like Piccadilly Circus. The traffic nearest the centre moves fastest; that further out more slowly, while that at the extreme edge merely crawls – at least by comparison with the fast traffic near the centre. . . .

Before we leave Piccadilly Circus, it should be understood that we cannot represent the solar system by putting up a statue of Eros in the middle to represent the sun, and letting nine taxicabs gyrate round it to represent the nine planets. The statue is far too big to represent the sun, and the taxicabs are enormously too big to represent planets. If we want to make a model to scale, we must take a very tiny object, such as a pea, to represent the sun. On the same scale the nine planets will be small seeds, grains of sand and specks of dust. Even so, Piccadilly Circus is only just big enough to contain the orbit of Pluto, the outermost planet of all. Think of a pea and nine tiny seeds, grains of sand and specks of dust in Piccadilly Circus, and we see that the solar system consists mainly of empty space. It is easy to understand why the planets look such tiny objects in the sky.

Yet the solar system is very crowded compared with most of space. If a pea and nine smaller objects in Piccadilly Circus represent the sun and planets, the nearest of the stars will be represented by a small seed somewhere near Birmingham – all in between is empty space.

JAMES JEANS, *The Stars in their Courses*, 1931

II. The Atom

Before we pass on, let us consider the general picture which we

---

[1] *Jeans's theory as to the origin of the solar system is no longer thought to be the most likely explanation, but the traffic is none the less one-way. (Ed.)*

must now form of an atom. At the centre we have a nucleus which, if we magnify it a million million times, will be about the size of a pea. It must not, of course, be thought of as a definite body like a small pea, but rather as a centre of electric force pervading the surrounding space. What we mean by the size is that, when we approach from outside the vague boundary which takes the place of a surface, the force suddenly becomes very great. If we had a very strong fire nobody would approach nearer to it than a certain distance: this would mark out a kind of limit, though there would be no actual wall there. It is somewhat in this sense that we speak of a boundary and a size for the nucleus.

Around this nucleus clusters a swarm of electrons, some on the whole closer, others farther out, the number and arrangement depending on the kind of atom. In the case of the heaviest atoms there are over ninety of them. These electrons occupy a space which, when magnified like the nucleus a million million times, would be about a hundred yards across. They occupy it not as water occupies a tank, for in the case of hydrogen there is but one of them, but in the sense that they patrol it like a guard. They prevent the electron patrol of other atoms from entering, and this patrolled sphere we therefore call the size of the atom, since two atoms brought together by ordinary collisions will bounce off from one another as soon as these loosely occupied spaces touch.

E. N. DA C. ANDRADE, *An Approach to Modern Physics*, 2nd
   edition, 1959

Jeans speaks of making 'a model to scale', and scale, together with speed and direction of travel, is all his analogies set out to convey. Similarly with Andrade's scale model, though the barrier of heat and the guard patrolling an area are attempts to convey more difficult ideas. But in both cases the analogies have been thought up in order to express an already existing idea in more easily understood terms.

Compare this use of analogy with the use made by Huxley and Young in Chapter Three. All three analogies for the living body – machine, whirlpool, and population – have obviously been useful, and useful to others besides those two authors, as tools of thought as well as methods of expressing that thought.

III. The Adding Machine

In order to have some picture of how the brain works it is useful to think of it as a gigantic government office – an enormous ministry, whose one aim and object is to preserve intact the country for which it is responsible. Ten million telegraph wires bring information to the office, coded in dots. These correspond to the sensory or input fibres reaching the brain. In the office one must try to imagine nearly 15,000,000,000 clerks, that is to say more than six times more people than there are at present in the whole world.[1] They correspond to the cells of the brain, and we can imagine them sitting in closely packed rows, as the brain cells are arranged. Every clerk has a telephone and receives coded messages either from outside the office or from some other part of it. So each nerve-cell of the brain receives nerve-impulses, either from the sense organs or from other brain cells near to it or far away. Each clerk spends most of his time sending code messages on his telephone to some other group, which may be near or far. So every nerve-cell has an out-going fibre, which may be long or short. But the clerks can also influence their neighbours by whispering 'silence'; obviously if a group of them starts doing this then a wave of quiet will pass over that area and it will send out no messages for a while.

In this way most elaborate patterns of activity will grow up between the huge numbers of clerks throughout the building. There are circuits by which messages are sent from one department to another and then back to the first and so on indefinitely. Messages will go round and round, but be influenced by incoming messages and by the waves of silence. However, the whole office is so arranged that some of the telephones eventually transmit instructions to workers outside, directing them how to run the country and bring food, drink, and other necessities to the govern-

[1] *Five times as many would be nearer the mark today. (Ed.)*

ment office. So some of the brain cells carry impulses that control the actions of the muscles, especially those of the hands, tongue and lips.

How would this office be organized, how would it convert the information it received into orders for the governing of the territory it controls? Everything, surely, would depend on its having arrangements by which all relevant information could be brought together to produce the right answer to every question put to it. This is just what happens in the nervous system. The sense organs transmit information to whichever departments of the brain can use it. But how does the brain bring this information together, so as to send out the right orders to the territory it is responsible for – the body? To find this out we may return to the comparison of brains with calculating machines. Information reaches the brain in a kind of code, you remember, of impulses passing up the nerve-fibres. Information already received is stored in the brain either by sending impulses round closed circuits, or in some form corresponding to a print. This is just what calculating machines do – they both store old information and receive new information and questions in coded form. The information received in the past forms the machine's rules of action, coded and stored away for reference. When asked a question, it puts it into code, and, by a process that is essentially one of adding and subtracting very fast, the machine can then refer the question to the rules that are already stored in it, and so produce the right answer. Similarly the brain is constantly relating the new impulses that reach it to the information already stored away in its tissues. To show the closeness of the parallel, think of a machine that would act as a cricket umpire. The wicket-keeper has whipped off the bails and called 'How's that?' A camera, rather like a television camera, turns the sequence of events into a code of dots and dashes. The machine already contains the rules of cricket, also in a code of dots and dashes. The machine could now proceed to fit together the coded report of the situation and the rules and answer with the word 'Out'. No such machine has in fact yet been built – but it might be. Its action would depend on being able to fit together the input from the camera with the rules. Modern calculating machines can do this sort of thing because the code is all in a simple form like dots and dashes, actually o's and 1's, and therefore involves adding and subtraction, though an enormous number of calculations may be necessary to determine the answer.

The brain has an even greater number of cells than there are valves in a calculator and it is not at all impossible that it acts quite like an adding machine, in some ways. . . . However, we still do not know exactly how the brain stores its rules or how it compares the input with them. It may use principles different from those of these machines.

J. Z. YOUNG, *Doubt and Certainty in Science*, 1951 (Reith Lectures, 1950)

IV. The Enchanted Loom

A scheme of lines and nodal points, gathered together at one end into a great ravelled knot, the brain, and at the other trailing off to a sort of stalk, the spinal cord. Imagine activity in this shown by little points of light. Of these some stationary flash rhythmically, faster or slower. Others are travelling points, streaming in serial trains at various speeds. The rhythmic stationary lights lie at the nodes. The nodes are both goals whither converge, and junctions whence diverge, the lines of travelling lights. The lines and nodes where the lights are, do not remain, taken together, the same even a single moment. There are at any time nodes and lines where lights are not.

Suppose we choose the hour of deep sleep. Then only in some sparse and out of the way places are nodes flashing and trains of light-points running. Such places indicate local activity still in progress. At one such place we can watch the behaviour of a group of lights perhaps a myriad string. They are pursuing a mystic and recurrent manoeuvre as if of some incantational dance. They are superintending the beating of the heart and the state of the arteries so that while we sleep the circulation of the blood is what it should be. The great knotted headpiece of the whole sleeping system lies for the most part dark, and quite especially so the roof-brain.[1] Occasionally at places in it lighted points flash or move but soon subside. Such lighted points and moving trains of light are mainly far in the outskirts, and wink slowly and travel slowly. At intervals even a gush of sparks wells up and sends a train down the spinal cord, only to fail to arouse it. Where, however, the stalk joins the headpiece, there goes forward in a limited field a remarkable display. A dense constellation of some thousands of nodal points bursts out every few seconds into a short phase of rhythmical flashing. At first a few lights, then more, increasing in rate and number with a deliberate crescendo to a climax, then to decline and die away. After due pause the efflorescence is repeated. With each such rhythmic outburst goes a discharge of trains of travelling lights along the stalk and out of it altogether into a number of nerve-branches. What is this doing? It manages the taking of our breath the while we sleep.

Should we continue to watch the scheme we should observe after a time an impressive change which suddenly accrues. In the great head-end which has been mostly darkness spring up myriads of twinkling stationary lights and myriads of trains of moving lights of many different directions. It is as though activity from one of those local places which continued restless in the darkened main-mass suddenly spread far and wide and invaded all. The great topmost sheet of the mass, that where hardly a light had twinkled or moved, becomes now a sparkling field of rhythmic flashing points with trains of travelling sparks hurrying hither and thither. The brain is waking and with it the mind is returning. It is as if the Milky Way entered upon some cosmic dance. Swiftly the head-mass becomes an enchanted loom where millions of flashing shuttles weave a dissolving pattern, always a meaningful pattern though never an abiding one; a shifting harmony of sub-patterns. Now as the waking body rouses, sub-patterns of this great harmony of activity stretch down into the unlit tracks of the

[1] *That part of the brain where sense-impressions are received and conscious acts originate. (Ed.)*

stalk-piece of the scheme. Strings of flashing and travelling sparks engage the lengths of it. This means that the body is up and rises to meet its waking day.

CHARLES SHERRINGTON, *Man on his Nature*, 1940 (Gifford Lectures, Edinburgh, 1937–8)

Young uses first the analogy of the office, then that of the calculating machine, in much the same way as he used whirlpools and populations in Chapter Three, as means to arrive at and convey some understanding of two different aspects to the way the brain works. Sherrington uses something nearer to poetic imagery to give us an impression, a sense, an awareness of the nature of the brain at work.

The styles used by the two authors – choice of word, rhythm and structure of sentence – are as different as their aims.

v. Likenesses

Man has only one means to discovery, and that is to find *likenesses* between things. To him, two trees are like two shouts and like two parents, and on this likeness he has built all mathematics. A lizard is like a bat and like a man, and on such likenesses he has built the theory of evolution and all biology. A gas behaves like a jostle of billiard balls, and on this and kindred likenesses rests much of our atomic picture of matter.

In looking for intelligibility in the world, we look for unity; and we find this (in the arts as well as in science) in its unexpected likenesses. This indeed is man's creative gift, to find or make a likeness where none was seen before – a likeness between mass and energy, a link between time and space, an echo of all our fears in the passion of Othello.

So, when we say that we can explain a process, we mean that we have mapped it in the likeness of another process which we know to work. We say that a metal crystal stretches because its layers slide over one another like cards in a pack, and then that some polyester yarns stretch and harden like a metal crystal.

That is, we take from the world round us a few models of structure and process (the particle, the wave, and so on), and when we research into nature, we try to fit her with these models.

Yet one powerful procedure in research, we know, is to break down complex events into simpler parts. Are we not looking for the understanding of nature in these? When we probe below the surface of things, are we not trying, step by step, to reach her ultimate and fundamental constituents?

We do indeed find it helpful to work piecemeal. We take a sequence of events or an assembly to pieces: we look for the steps in a chemical reaction, we carve up the study of an animal into organs and cells and smaller units within a cell. This is our atomic approach, which tries always to see in the variety of nature different assemblies from a few basic units. Our search is for simplicity, in that the distinct units shall be few, and all units of one kind identical.

And what distinguishes one assembly of these units from another? the elephant from the giraffe, or the right-handed molecule of sugar from the left-handed? The difference is in the organization of the units into the whole; the difference is in the structure. And the likenesses for which we look are also likenesses of structure.

This is the true purpose of the analytic method in science: to shift our gaze from the thing or event to its structure. We understand a process, we explain it, when we lay bare in it a structure which is like one we have met elsewhere.

> J. BRONOWSKI, *Science as Foresight* (In *What is Science?*), 1956

VI. 'Wrong Number'

These sensory systems, before anyone knew how they worked, used to be compared with telephone exchanges, simply because messages were known to travel between the end organs and the brain. Taste, least important of the senses, could then be conveniently described as a local exchange with very few outside wires and only four numbers of its own – sweet, sour, bitter and salt. But the analogy is just about as misleading as it could be. This must be explained.

In a telephone system the meaning of a message received depends on the sender; in a sensory system the meaning depends on the receiver. When nerve impulses travel from a sense organ, it is their destination on the cortex which determines, in the first place, the character of the sensation, not the sense organ from which they come. If, when you get a number on the telephone, you give a message, the message remains the same, even if you give it to a wrong number. The result of such an error in the brain is very different. Supposing some vinegar comes in contact with one of the sensitive end organs of taste in the tip of your tongue and 'gets a wrong number' – that is, say, supposing the nerve fibre conducting the impulse provoked by the vinegar, instead of connecting with its proper reception area, becomes in some way cut and grafted onto a nerve fibre leading from the ear to the brain – what do you think you would taste? You would taste nothing. You would hear a very loud and startling noise. Every time the nerve end in your tongue was stimulated you would have a similar hallucination. If, instead, one auditory nerve were in this way misconnected with an optic nerve, when you heard music you would see visions.

W. GREY WALTER, *The Living Brain*, 1953

VII. Wrong Analogy

Here are some examples to show the sort of thing that I mean. A most distinguished experimental physicist is talking about causality, and the principle of indeterminacy, due to Heisenberg, which appears to be related to it. After noticing that we can never measure the position and velocity of an electron (or any other particle either) with complete accuracy, so that we can never have any hard and rigid determinism of the kind that was fashionable seventy years ago, he asks the question: Are we therefore to abandon the idea of causality, or should we suppose that God takes control by means of causes that we shall never fully comprehend?

'It is for me easier', he writes, 'to suppose that there are causes that elude, and must for ever elude, our search, rather than to suppose that there are no causes at all. . . . In short we must admit causes beyond our comprehension. The electron leads us to the doorway of religion.'[1]

. . . This particular issue is an important one, and warrants a somewhat fuller description than some of my other illustrations, in view of the wide use which has been made of Heisenberg's uncertainty principle in ways like this. The problem hinges on what is implied by the statement that an electron has momentum (or velocity) and position. When we use language of this sort, we have in mind the picture of a small billiard ball which by analogy with larger billiard balls we ought therefore to be able to describe in terms of momentum and position. But why should an electron be like this? The plain truth is that we don't know. There is no finally convincing ground to justify us in calling it a particle at all. It is true that we find it exceedingly convenient for many purposes to treat it as if it really were a particle: but there are other occasions when it is far more convenient to treat it as if it were something quite different, about which we may use the language of waves and wavelengths and frequencies. In this second language position and momentum do not have the same meanings as when we are using the billiard ball picture. The uncertainty principle applies to both languages. In the particle language it tells us that there are limits to the precision of our measurement of position and velocity. In the wave language it speaks of limits to measurement of frequency and time. The uncertainty principle says nothing at all about whether we should use the one language or the other; i.e. whether there really is a particle with a position and a momentum. This is an undecidable question, and probably in the last resort it has no meaning to ask it. The clue to the whole problem is quite simple and straightforward once we realise that the uncertainty principle only talks about the results of measurement. It says

[1] Professor E. N. da C. Andrade, in *The Listener*, 10 July 1947.

nothing about the validity or otherwise of the model which we are using. Here the considered judgment of the physicists after twenty-five years thinking about it, is that the arbitrariness of a wave or a particle description warns us that we must enlarge our concepts. An electron is not a particle, though it may be good enough for many purposes to treat it as if it were. An electron is not a wave, though again for certain other purposes it may be convenient to treat it as if it were. This means that the electron does not lead us to the gateway of religion: it leads us to think a little more deeply about our science, and to modify our fundamental concepts to bring them into line with the increasing variety of our experiments. Once we admit that the electron need not be pictured as a tiny particle, the uncertainty relation has nothing more to say about freewill.

C. A. COULSON, *Science and Christian Belief*, 1955 (McNair Lectures, 1954)

Analogies can go wrong. None the less, both Grey Walter and Coulson are dependent on analogy, on likeness (even to the very lack of it), to say what they have to say and perhaps even to think what they have to think.

Lack of the right model or analogy or concept results in, indeed amounts to the same thing as, a failure of language. For conceptual thought results in and from, and is inseparable from, the use of language (though, of course, language can be used to express other things besides conceptual or abstract thought). And language, particularly when handling abstract matters, is itself rooted in likeness and draws on metaphor, dead or alive, in almost every sentence. (Witness Whitehead, pp. 92-3.)

What is to happen on those frontiers to human thought where no likenesses seem to exist? Does mathematics, the new language of physics, (a) function without the aid of likeness, or (b) discover or create new forms or kinds of likeness?

CHAPTER SEVEN

Cause and Effect or Blind Chance?

1. The Laws of Cause

More clearly than anyone else, it was the French thinkers of the Encyclopédie who drew the conclusion that all scientific prediction is like astronomical prediction. Given the whereabouts and the speeds of all the heavenly bodies at this instant, Newton had said, we can predict their movements from now to eternity. If that is so, said the French mathematician Laplace, then imagine yourself given the whereabouts and the speeds at this instant of every atom in the universe. Given all these, you can forecast the fate of the universe, its molecules and its men, its nebulae and its nations, from now into eternity. And more than this: you can go backward in time as well as forward, and reconstruct the past to eternity. Of course, the hope of actually carrying out such a calculation is rather fanciful. Nevertheless, science to Laplace remained the discovery of causal laws which help us to approach to this hope more and more nearly.

Laplace saw the lavish implications of this view clearly and he stated them boldly. This view has always been faced with some difficulties, particularly in finding a place for human action; and various dialectics and finesses have been invented to avoid them. It has been suggested, for example, that there are moments when the natural laws alter abruptly, and an increase of quantity tips over into a change of quality. But since it is still postulated that these critical steps are fully determined in time and in character by what has gone before, and that the new laws are caused by these changes, there is no real break in causality. There is an abrupt turn, but given all the facts Laplace asked for, even the turn is calculable.

These devices therefore do not deny a belief in complete and universal causality. They make the work of the computer more difficult, forward or backward in time. But they do not change its nature; it remains a purely mathematical task of solving some hypothetical array of equations of motion. Such devices complicate the belief in causality, but they do not change it, and in the last century they could not be allowed to change it.

There are several reasons why this belief will no longer do. The reasons are of different weight, and I myself am most impressed by a reason which is not conclusive, but which does shake my own confidence and must I think shake that of others. Here we have been thinking for nearly three hundred years that if there is one causal law which is certain beyond all challenge, it is the law of gravitation. The whole tradition of causality derives from its triumph. A hundred years ago, when the distant planet Uranus seemed not to be keeping time, we took it for granted that some unseen planet still further away must be disturbing it by its gravitational force. Two men, Adams in England and Leverrier in France, working with no knowledge of one another, and with nothing but pencil, paper and Newton's laws, calculated where such a planet must be. And when the great telescope at Berlin was turned to the spot, there was Neptune clear to the eye, and spectacular in its vindication of the unalterable laws of gravitation.

And yet, and yet, the laws of gravitation have gone. There is no gravitation; there is no force at all; the whole model was wrong. All that theory was no more than a happy approximation to what really happens. When Newton brought in force as a cause, he was giving to matter the human property of effort, as much as Aristotle once gave it human will. The true causes are now embedded in the nature of space and the way in which matter distorts space; and they have no resemblance to the causes in which we believed for nearly three hundred years. Ironically, Adams and Leverrier merely postponed the catastrophe by sixty years. For one beginning of the crisis in classical physics about 1900 was an oddity like the one they had set out to explain; only now it was the planet Mercury which was not keeping time. But search as we might, we could find no new Neptune to blame for the irregularity. It was cleared up only by a radical overhaul of the basic assumptions in Newton's philosophy, particularly in his conception of time.

I have said that this is not a final objection to causal laws. After all, the new theory which Einstein put in place of the old, although as a field theory it is less mechanical than Newton's, is still a causal theory. And Einstein, almost alone among the great physicists of today, continues stoutly to argue on behalf of causality. Yet it does seem to me, for two reasons, that this overthrow of a long accepted cause must deeply shake our confidence. For one thing, the whole conception of causes in science springs historically from the triumph of gravitation. And for another, we see now that it is possible to have every human faith in a causal mechanism, every assurance that this is how nature works, that here is her very action laid bare, and every demonstration that some apparent departure really fits in with that cause – we could have all these, unviolated and gaining in strength for two hundred years. And yet at the end we find that the cause was a fiction. Something else was at work, which has nothing in common with that famous cause. The machine never was a copy of nature. It was only a kind of gigantic planetarium, which got the heavenly bodies to the right place at the right time, but whose causal mechanism was no more like nature's than Ptolemy's itself.

Einstein found the flaw in Newton's theory of gravitation by looking into its very heart. There he found the assumption that time and space are given absolutely, and are alike for all observers. But when he thought through the steps by which different observers can actually compare their time in space, he found them at odds with this assumption. We cannot compare the time in two different places without sending a signal from one to the other, which must itself take time in its passage. As a result, Einstein showed that there is no universal 'now'; there is only 'here and now' for each observer, so that space and time are inextricably woven together, and are aspects of a single reality. Moreover, the structure of space in turn cannot be disentangled from the matter which is embedded in it.

In Einstein's Relativity physics then time is not a strict sequence of universal before and after. Closely spaced events which appear in one order to one observer may appear in the opposite order to another. Hume and John Stuart Mill had insisted long ago that the essence of cause and effect is their sequence: cause must come before and effect must follow after. Thus Einstein's new conception of time itself adds another difficulty to the definition of causality.

However, this difficulty also can be removed, and is not final. The final difficulty comes from another field, which is that of small scale or quantum physics. Einstein has made remarkable advances in this field also; indeed, he was given the Nobel Prize for his work not in Relativity but in quantum physics.

The fundamental step which created quantum physics had been taken in 1899 by Max Planck, when he discovered that energy, like matter, is not continuous, but appears always in packets or quanta of definite sizes. From the beginning, the ideas of quantum physics could not be matched with the classical mechanics of particles. Fantastic properties had to be given to an electron whenever it sent out or took up a quantum of energy. These difficulties grew until in the 1920's it began to be seen that we simply cannot make a theory to describe these minute happenings and still hope to keep it rigidly laid out in the classical pattern of causes and effects. There is no way at all of describing the present and future of these tiny particles and events so that they appear completely determined. This was put into a formal principle in 1927 by the German physicist Heisenberg, and given the sensible name of the principle of uncertainty.

Heisenberg showed that every description of nature contains some essential and irremovable uncertainty. For example, the more accurately we try to measure the position of a fundamental particle, of an electron say, the less certain will we be of its speed. The more accurately we try to estimate its speed, the more uncertain will we be of its precise position. Therefore we can never predict the future of the particle with complete certainty; because as a matter of fact we cannot be completely certain of its present. If we want to predict its future sensibly, then we must allow it to have some uncertainty: some range of alternatives, some slack – what engineers call some tolerance. We may have what metaphysical prejudices we choose, whether the future really and truly, essentially, is determined by the present. But the physical fact about these small scale events is beyond dispute. Their future cannot be foretold with complete assurance by anyone observing them in the present. And of course, once we have any uncertainty in prediction, in however small and distant corner of the world, then the future is essentially uncertain – although it may remain overwhelmingly probable.

I have said that this principle of uncertainty refers to very small particles and events. But these small events are not by any means unimportant. They are just the sorts of events which go on in the nerves and the brain and in the giant molecules which determine the qualities we inherit.

J. BRONOWSKI, *The Common Sense of Science*, 1951

Clearly we all act on the assumption that causes produce effects, just as we live our lives as if tables are solid and substantial. Equally clearly, from the trouble we take reaching decisions and the agony we go through regretting them, few of us subscribe to the determinism which a thoroughgoing philosophy of cause and effect entails. How helpful are Heisenberg's uncertainties in resolving the dilemma, and how urgent, or real, is the dilemma?

11. The Laws of Chance

We have to remember that no scientific method is truly deductive, taking the physical facts and reasoning from them conclusively to the laws which they follow. At the basis of scientific method lies the kind of imagination which Newton used, who defined a world of particles, postulated laws or axioms which these particles individually follow, and then showed that they combine to make a world much like that we know. Newton had no theory of what these ultimate particles are, and it is we who have tried to identify them first with molecules, then with atoms, then electrons and other minute and indivisible constituents of matter. We have failed. If the world is made of electrons and things like them, then it is certain that they do *not* behave like Newton's particles. They behave sometimes like waves and sometimes like particles; they do not have a precise place and speed at the same instant; and they have other oddities. And when we say for instance that whereabouts and speed cannot be exactly observed at the same time, we mean just this limitation: that we cannot make the hypothesis of individual particles and give them at the same time defined places and speeds in our equations.

These difficulties are not wholly to be traced to the search for cause and effect. Rather they arise because we have believed more deeply that all scientific happenings can be broken into smaller and smaller ultimate pieces, and that these pieces each obey causal laws. We have come to assume that any event which did not seem to flow from its antecedents of itself was sure to do so if we broke it into small enough pieces, either of fact or of matter. This analytic process has really been the basis of our notion of determinism. What we now see is that you cannot have both together. You cannot have a model made of minute particles and events, and have each particle and each event running on strictly causal orbits. Cause and effect are large-scale operations. But the analytic process in the end gives rise to a form of law which is different in type – a law of chance in place of cause. . . .

This is the method to which modern science is moving. It uses no principle but that of forecasting with as much assurance as possible, but with no more than is possible. That is, it idealises the future from the outset, not as completely determined, but as determined within a defined area of uncertainty. Let me illustrate the kind of uncertainty. We know that the children of two blue-eyed parents will certainly have blue eyes; at least, no exception has ever been found. By contrast, we cannot be certain that all the children of two brown-eyed parents will have brown eyes. And we cannot be certain of it even if they have already had ten children with brown eyes. The reason is that we can never discount a run of luck of the kind which Dr Johnson once observed when a friend of his was breeding horses. 'He has had', said

Dr Johnson, 'sixteen fillies without one colt, which is an accident beyond all computation of chances.' But what we can do is to compute the *odds* against such a run; this is not as hard as Johnson supposed. . . .

This area of uncertainty shrinks very quickly in its proportion if we make our forecasts not about one family but about many. I do not know whether this or that couple will have a child next year; I do not even know whether I shall. But it is easy to estimate the number of children who will be born to the whole population, and to give limits of uncertainty to our estimate. The motives which lead to marriage, the trifles which cause a car to crash, the chanciness of today's sunshine or tomorrow's egg, are local, private and incalculable. Yet, as Kant saw long ago, their totals over the country in a year are remarkably steady; and even their ranges of uncertainty can be predicted.

This is the revolutionary thought in modern science. It replaces the concept of the *inevitable effect* by that of the *probable trend*. Its technique is to separate so far as possible the steady trend from local fluctuations. The less the trend has been overlaid by fluctuations in the past, the greater is the confidence with which we look along the trend into the future. We are not isolating a cause. We are tracing a pattern of nature in its whole setting. We are aware of the uncertainties which that large, flexible setting induces in our pattern. But the world cannot be isolated from itself: the uncertainty *is* the world. The future does not already exist; it can only be predicted. We must be content to map the places into which it may move, and to assign a greater or less likelihood to this or that of its areas of uncertainty.

These are the ideas of chance in science today. They are new ideas: they give chance a kind of order; they re-create it as the life within reality. These ideas have come to science from many sources. Some were invented by Renaissance brokers; some by seventeenth century gamblers; some by mathematicians who were interested in aiming-errors and in the flow of gases and more recently in radio-activity. The most fruitful have come from biology within little more than the last fifty years. I need not stress again how successful they have been in the last few years, for example in physics: Nagasaki is a monument to that. But we have not yet begun to feel their importance outside science altogether. For example, they make it plain that problems like Free Will or Determinism are simply misunderstandings of history. History is neither determined nor random. At any moment, it moves forward into an area whose general shape is known but whose boundaries are uncertain in a calculable way. A society moves under material pressure like a stream of gas; and on the average, its individuals obey the pressure; but at any instant, any individual may, like an atom of the gas, be moving across or against the stream. The will on the one hand and the compulsion on the other exist and play within these boundaries. In these ideas, the concept of chance has lost its old dry pointlessness and has taken on a new depth and power; it has come to life. Some of these new ideas have begun to influence the arts: they can be met vaguely in the novels of the young French writers. In time they will liberate our literature from the pessimism which comes from our divided loyalties: our reverence for machines and, at odds with it, our nostalgia for personality. I am young enough to believe that this union, the union as it were of chance with fate, will give us all a new optimism.

J. BRONOWSKI, *The Common Sense of Science*, 1951

CHAPTER EIGHT

Evolution and Man

1. Naturally Selected

One somewhat curious fact emerges from a survey of biological progress as culminating for the evolutionary moment in the dominance of *Homo sapiens*. It could apparently have pursued no other general course than that which it has historically followed: or, if it be impossible to uphold such a sweeping and universal negative, we may at least say that among the actual inhabitants of the earth, past and present, no other lines could have been taken which would have produced speech and conceptual thought, the features that form the basis for man's biological dominance.

Multicellular organization was necessary to achieve the basis for adequate size: without triploblastic development and a blood-system, elaborate organization and further size would have been impossible. Among the coelomates, only the vertebrates were eligible as agents for unlimited progress, for only they were able to achieve the combination of active efficiency, size, and terrestrial existence on which the later stages of progress were of necessity based. Only in the water have the molluscs achieved any great advance. The arthropods are not only hampered by their necessity for moulting; but their land representatives, as was first pointed out by Krogh, are restricted by their tracheal respiration to very small size. They are therefore also restricted to cold-bloodedness[1] and to a reliance on instinctive behaviour. Lungs were one needful precursor of intelligence. Warm blood was another, since only with a constant internal environment could the brain achieve stability and regularity for its finer functions. This limits us to birds and mammals as bearers of the torch of progress. But birds

[1] *The smaller the body, the greater its surface area in relation to its volume, the more effective its means of losing heat to or absorbing heat from its surroundings, and the less effective any internal mechanism for regulating its temperature. (Ed.)*

were ruled out by their depriving themselves of potential hands in favour of actual wings, and perhaps also by the restriction of their size made necessary in the interests of flight.

Remain the mammals. During the Tertiary epoch, most mammalian lines cut themselves off from the possibility of ultimate progress by concentrating on immediate specialization. A horse or a lion is armoured against progress by the very efficiency of its limbs and teeth and sense of smell: it is a limited piece of organic machinery. As Elliot Smith has so fully set forth, the penultimate steps in the development of our human intelligence could never have been taken except in arboreal ancestors, in whom the forelimb could be converted into a hand, and sight inevitably became the dominant sense in place of smell. But, for the ultimate step, it was necessary for the anthropoid to descend from the trees before he could become man. This meant the final liberation of the hand, and also placed the evolving creature in a more varied environment, in which a higher premium was placed upon intelligence. Further, the foetalization necessary for a prolonged period of learning could only have occurred in a monotocous species.[1] Weidenreich (1941) maintains that the attainment of the erect posture was a necessary prerequisite for the final stages in human cerebral evolution.

The last step yet taken in evolutionary progress, and the only one to hold out the promise of unlimited (or indeed of any further) progress in the evolutionary future, is the degree of intelligence which involves true speech and conceptual thought: and it is found exclusively in man. This, however, could only arise in a monotocous mammal of terrestrial habit, but arboreal for most of its mammalian ancestry. All other known groups of animals, except the ancestral line of this kind of mammal, are ruled out. Conceptual thought is not merely found exclusively in man: it could not have been evolved on earth except in man.

---

[1] *If, normally, only one embryo is fertilized at a time, its rate of development in the womb (since it has no rivals to compete with) can be slow. This slow rate of growth continues after birth, allowing a long period for learning before the limiting factor of maturity is reached. (Ed.)*

Evolution is thus seen as a series of blind alleys. Some are extremely short – those leading to new genera and species that either remain stable or become extinct. Others are longer – the lines of adaptive radiation within a group such as a class or sub-class, which run for tens of millions of years before coming up against their terminal blank wall. Others are still longer – the lines that have in the past led to the development of the major phyla and their highest representatives; their course is to be reckoned not in tens but in hundreds of millions of years. But all in the long run have terminated blindly. That of the echinoderms, for instance, reached its climax before the end of the Mesozoic. For the arthropods, represented by their highest group, the insects, the full stop seems to have come in the early Cenozoic: even the ants and bees have made no advance since the Oligocene. For the birds, the Miocene marked the end; for the mammals, the Pliocene.

Only along one single line is progress and its future possibility being continued – the line of man. If man were wiped out, it is in the highest degree improbable that the step to conceptual thought would ever again be taken, even by his nearest kin. In the ten or twenty million years since his ancestral stock branched off from the rest of the anthropoids, these relatives of his have been forced into their own lines of specialization, and have quite left behind them that more generalized stage from which a conscious thinking creature could develop. Although the reversibility of evolution is not an impossibility *per se*, it is probably an actual impossibility in a world of competing types. Man might conceivably cause the capacity for speech and thought to develop by long and intensive selection in the progeny of chimpanzees or gorillas; but Nature, it seems certain, could never do so.

One of the concomitants of organic progress has been the progressive cutting down of the possible modes of further progress, until now, after a thousand or fifteen hundred million years of evolution, progress hangs on but a single thread. That thread is the human germ-plasm. As Villiers de l'Isle-Adam wrote in *L'Ève Future*, 'L'Homme . . . seul, dans l'univers, n'est pas fini.'

JULIAN HUXLEY, *Evolution: the Modern Synthesis*, 1942

Huxley's two main theses, that efficiency, when it results from over-specialization, can be a bar to further progress, and that any significant evolutionary advances yet to be made on earth must depend on man, are developed at greater length and in highly individual manners by the authors of the remaining two extracts in this chapter.

Whether it is Rostand or his translator who should be credited with 'whittled down and pulverized' and 'dissolve into a mere cog in the great machine' I have been unable to discover. The most distinctive, and in many ways most helpful, feature of Medawar's style is that, like some teaching machines, it will not let the learner proceed to the next stage without making sure by quick revision that all previous stages have been mastered. Such periodic 'taking stock' is often of as great assistance to writer as to reader.

11. Good at Nothing, Good at Everything

According to the philosopher Renouvier, unity and diversity are the two poles of all existence, and for Spencer[1] the transition from unity to diversity was the fundamental law of all progress. Starting with biological ideas derived from German physiology, he maintained that the development of the earth, of society, of governments, of industry, of commerce, of language, of literature, of science and of art followed the same evolution from the simple to the complex that is characteristic of the development of all living beings and the evolution of all species.

Examples: society began as a collection of individuals with the same rights and obligations; every man was originally warrior, hunter, fisherman and worker all at once. Then came specialization and with it social distinctions. We know to what extent the division of labour is a reality in our own society, and recall that the very term was originally borrowed by Adam Smith from biology. Durkheim wrote that 'not only are the tasks within a given factory highly specialized but each factory in turn is itself a specialized unit. . . . The division of labour is not restricted

---

[1] *Herbert Spencer, 1820–1903. (Ed.)*

to the economic world, its growing influence can be observed in the most varied social spheres. Political, administrative and judicial functions are becoming more and more specialized. The same is true of artistic and scientific work also. We have gone a long way since the time when philosophy was the only science; it has split into a multitude of special disciplines, each with objects, methods and approaches of its own.'

Auguste Compte also complained about this fragmentation of science. What would he say today, when every discipline is whittled down and pulverized to the point where no man of science can pretend to grasp more than a small portion of even his own branch of knowledge? So rapid is this drive towards specialization, that its effects can be felt within the short span of one life.

To take a personal example, it is no longer possible, as it was in my youth, to claim knowledge of biology: we merely study some aspect of genetics, embryology, immunology, histology, ecology, or endocrinology. It is because of this increase in '-ologies' that team-work in research is becoming more and more essential. . . .

This irresistible march from unity to diversity is not without drawbacks or even dangers. Economists have long deplored the sad consequences of the division of labour which, by leading to the performance of monotonous and restricted tasks, robs work of all its personal meaning. Marx accused mechanization of dismembering man and the reader will recall Jean-Baptiste Say's bitter reflections on the workman whose sole task it was to turn out the eighteenth part of a pin.

Carrel, for his part, strongly denounced the spiritual dangers arising from the departmentalization and fragmentation of knowledge. Deploring the fact that knowledge can no longer exist entire in any one brain, he expressed the hope that some men would devote their lives to gathering knowledge as such, thus safeguarding its unity and integrity. In the medical field, many specialists have ceased to look upon their patients as whole men, considering them instead as so many organs, if not parts of organs, with a consequent restriction of judgement that may border on fetishism. All illnesses are explained away as pituitary disorders or as related to the proportion of cholesterol in the blood.

Finally, as pedagogues have realized long ago, specialization may be a grave danger to education since only a wide cultural background can lay the intellectual foundations for fruitful inquiry. Some, it is true, have maintained that specialization in early life is a *sine qua non* of later skill and efficiency. Admittedly, as science and technology are becoming more and more specialized, it is perhaps unavoidable that the individual should follow suit, but is he to dissolve into a mere cog in the great machine, a mere cell in the great organism?

Now that we have stated the problem, let us examine its relevance to biology. To do so, we must shelve what doctrinal or sentimental preferences we may hold, the better to look at the positive facts. We shall begin by considering the transition from unity to differentiation in the development of the human being.

A human being is a colony of different types of cells, each type consisting of millions of individuals. Nerve cells, bone cells, blood cells, epithelial cells, gland cells, muscle cells, etc., are just so many characteristic and stable types differing in form, size, function, lifetime and behaviour. . . .

All animals arise from a single cell, the ovum, which, in turn, is formed by the union of two germ cells: the maternal ovule and the paternal spermatozoon. The nucleus of the human ovum contains 48 chromosomes, 24 of which are derived from the father, and 24 from the mother. The development of the human being is the result of a sequence of cell divisions: the ovum dividing into two cells, then into four, eight, etc. As the cell divides, so do the chromosomes, so that, accidents excepted, every cell of the body receives the 48 chromosomes of the ovum.

Since the human ovum weighs less than a thousandth of a milligram, while the new-born baby weighs roughly seven pounds, cell divisions clearly involve cell growth. Nor is that the whole story. As we have seen, an organism is not a colony of similar cells. In other words, at some stage in the development, a process of *cellular differentiation* must have taken place. . . .

The egg of the *Triton*[1] is a small ovoid of brownish protoplasm surrounded by a transparent membrane. The female fertilizes the egg externally with a supply of semen previously supplied by the male during their graceful love-play. Some time later, this egg, like all fertilized eggs, begins to divide. In the course of roughly one day – the time depending on the temperature of the water – the egg has divided into a large number of cells; it now resembles a small mulberry, and is therefore called a morula (Latin *morum* = mulberry).

The morula acquires a central cavity and becomes the blastula, which, in turn, develops into the gastrula. Now, from his studies of the morula, the biologist – or rather, the embryologist, for we must respect specialization! – knows by what processes this small multi-cellular sphere has become transformed into a full-grown vertebrate. He knows these processes in such detail that he can predict with certainty what part of the future animal will arise from what particular area or zone of the morula. . . .

If, during the early stages of the embryo's life, a piece of presumptive epidermis is grafted into the neural plate region, it will behave like nervous tissue, while a piece of presumptive neural plate develops in epidermal regions into typical epidermis. In other words, the development of the embryo is in no way affected by the substitution. It follows its normal course: the skin cells becoming nerve cells, and the nerve cells becoming skin cells. At this stage, the cells are still identical; they are not yet differentiated – or if they are, at least not yet to the point where they can no longer change their destiny by changing position.

Thus, all the cells of the young embryo are interchangeable. They are said to be 'totipotent', i.e. they are jack-of-all-trade cells. We are still at the hour of unity, the hour of specialization has not yet struck. The future of the cell still depends on its position, its fate can be changed by simple displacement. Its potential future far exceeds its normal future.

Now let us repeat our experiment with another *Triton* egg,

[1] *A small newt, on whose embryo the experiment described was first carried out by Hans Spemann. (Ed.)*

older by some hours (late gastrula stage). This time the result is quite different.

Epidermal cells remain epidermal cells and nerve cells remain nerve cells, no matter to what region they are transplanted. In other words, at this slightly later stage in the embryo's life, differentiation has begun to play an irrevocable and irreversible role. Distinctions between cells have become definite and can no longer be broken down by any manipulations. Their future is settled once for all. No longer jack-of-all-trade cells, their potential future no longer exceeds their real future. . . .

Though our somatic cells are the result of early cell differentiation, our reproductive cells, the so-called germ-line, are not. Biologists have long been discussing whether this distinction between somatic and germ cells is, in fact, fundamental. Without taking sides in this specialist dispute, we are on safe ground in saying that everything happens *as if* this were so. In many animals the separation of a special group of cells, the germ-line determinants, becomes apparent from the very beginning of the animal's development. Sometimes an 'island' of special protoplasm can be made out in even the undivided egg. If this little island is destroyed, the egg will produce animals totally devoid of germ cells. The excision is, in fact, an early form of castration, castration *ab ovo* as it were. . . .

The egg, an undifferentiated cell, can therefore produce two types of cell, undifferentiated cells like itself which will form the germs, and differentiated cells which will form the body. Germ cells are never the result of the specialization of other cells – their lack of differentiation is continuous from one generation to the next.[1] Non-differentiation may be considered a form of idle capital, put aside for the animal's future. . . .

[1] *The author may mean no more than that, in the development of the body from embryo to adult, no differentiated cell ever gives rise by cell-division to an undifferentiated cell. It is also true that there is no known mechanism whereby, as is presupposed by the Lamarckian heresy, differentiated or somatic cells which alter in some way during the adult's life (e.g. those cells composing the more-than-usually-stretched necks of ancestral giraffes) can affect the structure of the germ cells in such a way that the alteration be passed on to the next generation. (Ed.)*

Having examined both the differentiation of cells and also the differentiation of individuals within a given species,[1] we shall now consider our last problem: the differentiation of species. For is not the species itself the result of differentiation within a given genus, and the genus the result of differentiation within a given family?

According to the classical theory of evolution, a common stock may produce a number of branches which, in time, become more and more distinct from one another. This evolutionary differentiation – the mechanism of which is not known, though it is certainly connected with the inherited cell nucleus – almost invariably goes hand in hand with *specialization*, and an example will explain the meaning of that term.

There is no doubt that primitive mammals had a hand with five fingers, vaguely resembling that of our frog. Now this primitive hand must have given rise to the claws of the carnivore, to the hoof of the horse, to the flipper of the seal, the wing of the bat. All these 'hands' go with a certain mode of life and with certain habits, and are said to be specialized forms of the original and general type in which all subsequent forms must have been contained.

Here we have a process that strongly resembles the development and differentiation of the various types of cells.

In fact, the development of an individual has often been compared to the development of the entire animal kingdom and, even though the analogy may be a little superficial, it is not without value.

To some extent, specialization and differentiation represent a kind of evolutionary impasse, for nothing new can be born of an over-specialized type. Now, in the past, the major advances in life have been wrought by a succession of 'generalized' undifferentiated

---

[1] 'This type of differentiation', *says the author in part of the passage omitted*, 'is characterized by the intolerance of the cells of one individual towards the cells of another. Moreover, this intolerance increases with age, two adults behaving as if they were even more incompatible than two youths and, *a fortiori*, than two embryos.'

types and such types were called 'soft' or plastic by the philosopher Leroy, to emphasize the fact that they played the same
part in the development of the species as the embryo plays in the
development of the individual.

To revert to the example of the hand, man – the crown of
evolution – has preserved the hand of the primitive mammal. This
is a clear sign that, in the line from which man has sprung,
specialization has never gone too far. In many other respects, too,
man has retained marked traits of the primitive mammal. As
Leroy puts it, man has 'preserved with striking freshness certain
zoological traits of the most ancient mammals known. His limbs,
number of digits, and arrangement of teeth are strangely primitive.'

Not being specially adapted either to swimming, or to running,
or to flying, not being good for any one thing man is good for
everything. Compared with quaint animals like the giraffe, the bat,
or the seal, man seems to be endowed with a pure, classical
structure, so much so that we might call him *the* mammal, the
king of mammals. It has been said of man that he is the only
mammal that could be called an embryo. . . .

We have seen that it is the absence of differentiation in germ
cells, which leads to the reproduction of individuals. . . . Only
because an egg is an undifferentiated cell, can it reproduce a
complete animal, unlike, say, a hepatic cell which can never
produce any but hepatic cells, a cartilage cell which can only
produce cartilage cells, or a blood cell which gives rise to none
but blood cells. Similarly, in the evolution of species, a highly
specialized form is a final and fixed form of which little or nothing
further can be expected. It is the undifferentiated forms that
contain all future potentials. We could say that, by and large,
*actual progress depends on differentiation, while potential progress is
possible only by virtue of non-differentiation.*

Non-differentiation is a kind of insurance against exhaustion,
senility and mechanization; it is like a reservoir of freshness, full
of promise for the future.

Leroy thought that, in this respect, the phenomena of life
could be compared to those of thought. He remarked on the

similarity between non-differentiation and invention on the one hand, and between differentiation and automatism on the other: 'Institutions, societies, codes of behaviour or philosophic systems always follow the same evolutionary rhythm: water-tight explanations that are the triumphs of one epoch, reveal their impotence when confronted with the thoughts of another; the hardening of habits provokes a lethal form of ankylosis, and while the promise of the future belongs to the inventive mind, hypertrophy extinguishes the virtues of initiative, and causes our institutions to be swamped by scholasticism.' And since philosophy was a less rigidly circumscribed discipline than science, and hence a less differentiated form of thinking, Leroy looked upon philosophy as the handmaiden of future progress and spiritual revival.

. . . Should individuals, groups, schools of thought or philosophy, and political parties be encouraged to cater for non-differentiation, plasticity and immaturity?

This conclusion might be a little too narrow, too rigid, too differentiated. . . . And, moreover, it is by no means certain that, while profiting by differentiation, we can deliberately cultivate non-differentiation by its side. Only nature may have her cake and eat it. . . .

What does seem certain is that individuals, groups, or schools of thought, will not evolve or advance, change or adapt themselves unless they contain undifferentiated elements, the buds of all future growth, and unless they retain the secret of youth in heart and mind.

'*La Beauté de l'enfance est de ne pas finir.*'

JEAN ROSTAND, *Error and Deception in Science*, 1960 (tr. A. J. Pomerans)

III. The New Evolution

In the last lecture, I shall discuss the origin in human beings of a new, a non-genetical, system of heredity and evolution based upon certain properties and activities of the brain. The existence of this non-genetical system of heredity is something you are perfectly well aware of. It was not biologists who first revealed to an incredulous world that human beings have brains; that having brains makes a lot of difference; and that a man may influence posterity by other than genetic means. Yet much of what I have read in the writings of biologists seems to say no more than this. I feel a biologist should contribute something towards our *understanding* of the distant origins of human tradition and behaviour, and this is what I shall now attempt. The attempt must be based upon hard thinking, as opposed to soft thinking; I mean, it must be thinking that covers ground and is based upon particulars, as opposed to that which finds its outlet in the mopings or exaltations of poetistic prose.

It will make my argument clearer if I build it upon an analogy. I should like you to consider an important difference between a juke-box and a gramophone – or, if you like, between a barrel organ and a tape-recorder. A juke-box is an instrument which contains one or more gramophone records, one of which will play whatever is recorded upon it if a particular button is pressed. The act of pressing the button I shall describe as the 'stimulus'. The stimulus is specific: to each button there corresponds one record, and *vice versa*, so that there is a one-to-one relationship between stimulus and response. By pressing a button – any button – I am, in a sense, instructing the juke-box to play music; by pressing this button and not that, I am instructing it to play one piece of music and not another. But – I am not giving the juke-box *musical* instructions. The musical instructions are inscribed upon records that are part of the juke-box, not part of its environment: what a juke-box or barrel-organ can play on any one occasion depends upon structural or inbuilt properties of its own. I shall follow Professor Joshua Lederberg in using the word 'elective' to describe the relationship between what the juke-box plays and the stimulus that impinges upon it from the outside world.

Now contrast this with a gramophone or any other reproducing apparatus. I have a gramophone, and one or more records somewhere in the environment outside it. To hear a particular piece of music, I go through certain motions with switches, and put a gramophone record on. As with the juke-box I am, in a sense, instructing the gramophone to play music, and a particular piece of music. But I am doing more than that: I am giving it musical instructions, inscribed in the grooves of the record I make it play. The gramophone itself contains no source of musical information; it is the record that contains the information, but the record reached the gramophone from the outside world. My relationship to the gramophone – again following Lederberg – I shall describe as 'instructive'; for, in a sense, I *taught* it what to play. With the juke-box, then – and the same goes for a musical-box or barrel-organ – the musical instructions are part of the system that responds to stimuli, and the stimuli are elective: they draw upon the inbuilt capabilities of the instrument. With a gramophone, and still more obviously with a tape recorder, the stimuli are instructive: they endow it with musical capabilities; they import into it musical information from the world outside.

It is we ourselves who have made juke-boxes and gramophones, and who decide what, if anything, they are to play. These facts are irrelevant to the analogy I have in mind, and can be forgotten from now on. Consider only the organism on the one hand – juke-box or gramophone; and, on the other hand, stimuli which impinge upon that organism from the world about it.

During the past ten years, biologists have come to realize that, by and large, organisms are very much more like juke-boxes than gramophones. Most of those reactions of organisms which we were formerly content to regard as instructive are in fact elective. The instructions an organism contains are not musical instructions inscribed in the grooves of a gramophone record, but *genetical* instructions embodied in chromosomes and nucleic acids. Let me give examples of what I mean.

The oldest example, and the most familiar, concerns the change that comes over a population of organisms when it undergoes an evolution. How should we classify the environmental stimuli that cause organisms to evolve? The Lamarckian theory, the theory that acquired characters can be inherited, is in its most general form, an *instructive* theory of evolution. It declares that the environment can somehow issue genetical instructions to living organisms – instructions which, duly assimilated, can be passed on from one generation to the next. The blacksmith who is usually called upon to testify on these occasions gets mightily strong arms from forging; somehow this affects the cells that manufacture his spermatozoa, so that his children start life specially well able to develop strong arms.[1] I have no time to explain our tremendous psychological inducement to believe in an instructive or Lamarckian theory of evolution, though in a somewhat more sophisticated form than this. I shall only say that every analysis of what has appeared to be a Lamarckian style of heredity has shown it to be *non*-Lamarckian. So far as we know, the relationship between organism and environment in the evolutionary process is an elective relationship. The environment does *not* imprint genetical instructions upon living things.[2]

---

[1] *A. R. Wallace, in his address given jointly with one by Darwin to the Linnean Society, London, on 1 July 1858, has this to say.* 'The hypothesis of Lamarck – that progressive changes in species have been produced by the attempts of animals to increase the development of their organs, and thus modify their structure and habits – has been repeatedly and easily refuted by all writers on the subject of varieties and species . . . but the view here developed renders such an hypothesis quite unnecessary, by showing that similar results must be produced by the action of principles constantly at work in nature. The powerful retractile talons of the falcon and the cat tribes have not been produced or increased by the volition of those animals; but among the different varieties which occurred in the earlier and less highly organized forms of these groups,* those always survived longest which had the greatest facilities for seizing their prey. *Neither did the giraffe acquire its long neck by desiring to reach the foliage of the more lofty shrubs, and constantly stretching its neck for the purpose, but because any varieties which occurred among its antetypes with a longer neck than usual* at once secured a fresh range of pasture over the same ground as their shorter-necked companions, and on the first scarcity of food were thereby enabled to outlive them. *Even the peculiar colours of many animals, especially insects, so closely resembling the soil or the leaves or the trunks on which they habitually reside, are explained on the same principle; for though in the course of ages varieties of many tints may have occurred,* yet those races having colours best adapted to concealment from their enemies would inevitably survive the longest.' (*Ed.*)

[2] *Recent work in America is beginning to suggest that as categoric a statement as this may not be entirely true. (Ed.)*

Another example: bacteriologists have known for years that if bacteria are forced to live upon some new unfamiliar kind of food-stuff or are exposed to the action of an anti-bacterial drug, they acquire the ability to make use of that new food, or to make the drug harmless to them by breaking it down. The treatment was at one time referred to as the *training* of bacteria – with the clear implication that the new food or drug *taught* the bacteria how to manufacture the new ferments upon which their new behaviour depends. But it turns out that the process of training belies its name: it is not instructive. A bacterium can synthesize only those ferments which it is genetically entitled to synthesize. The process of training merely brings out or exploits or develops an innate potentiality of the bacterial population, a potentiality under-written or subsidized by the particular genetic make-up of one or another of its members.[1]

The same argument probably applies to what goes on when animals develop. At one time there was great argument between 'preformationists' and those who believed in epigenesis. The pre-formationists declared that all development was an unfolding of something already there; the older extremists, whom we now laugh at, believed that a sperm was simply a miniature man. The doctrine of epigenesis, in an equally extreme form, declared that all organisms begin in a homogeneous state, with no apparent or actual structure; and that the embryo is molded into its adult form solely by stimuli impinging upon it from outside. The truth lies somewhere between these two extreme conceptions. The genetic instructions are preformed, in the sense that they are already there, but their fulfilment is epigenetic – an interpretation that comes close to an elective theory of embryonic development. The environment brings out potentialities present in the embryo in a way which (as with the buttons on a juke-box) is exact and discriminating and specific; but it does not *instruct* the developing

---

[1] *Given this innate ability of certain members of a bacterial population to cope with and survive new conditions fatal to most of their fellows, natural selection as outlined above by Wallace will speedily ensure the spread of such immunity to an increasing proportion of the total population. (Ed.)*

embryo in the manufacture of its particular ferments or proteins or whatever else it is made of. Those instructions are already embodied in the embryo: the environment causes them to be carried out.

Until a year or two ago we all felt sure that *one* kind of behaviour indulged in by higher organisms did indeed depend upon the environment as a teacher or instructor. The entry or injection of a foreign substance into the tissues of an animal brings about an immunological reaction. The organism manufactures a specific protein, an 'antibody', which reacts upon the foreign substance, often in such a way as to prevent its doing harm. The formation of antibodies has a great deal to do with resistance to infectious disease. The relationship between a foreign substance and the particular antibody it evokes is exquisitely discriminating and specific; one human being can manufacture hundreds – conceivably thousands – of distinguishable antibodies, even against substances which have only recently been invented, like some of the synthetic chemicals used in industry or in the home. Is the reaction instructive or elective? – *surely*, we all felt, instructive. The organism learns from the chemical pattern of the invading substance just how a particular antibody should be assembled in an appropriate and distinctive way. Self-evident though this interpretation seems, many students of the matter are beginning to doubt it. They hold that the process of forming antibodies is probably elective in character. The information which directs the synthesis of particular antibodies is part of the inbuilt genetical information of the cells that make them; the intruding foreign substance exploits that information and brings it out. It is the juke-box over again. I believe this theory is somewhere near the right one, though I do not accept some of the special constructions that have been put upon it.

So in spite of all will to believe otherwise, and for all that it seems to go against common sense, the picture we are forming of the organism is a juke-box picture – a juke-box containing genetical instructions inscribed upon chromosomes and nucleic acids in much the same kind of way as musical instructions are inscribed upon gramophone records. But what a triumph it would be if an organism could accept information from the environment – if the environment could be made to act in an instructive, not merely an elective, way! A few hundred million years ago a knowing visitor from another universe might have said: 'It's a splendid idea, and I see the point of it perfectly: it would solve – or could solve – the problems of adaptation, and make it possible for organisms to evolve in a much more efficient way than by natural selection. But it's far too difficult: it simply can't be done.

But you know that it has been done, and that there is just one organ which can accept instruction from the environment: the brain. We know very little about it, but that in itself is evidence of how immensely complicated it is. The evolution of a brain was a feat of fantastic difficulty – the most spectacular enterprise since the origin of life itself. Yet the brain began, I suppose, as a device for responding to elective stimuli. *Instinctive* behaviour is behaviour in which the environment acts electively. If male sex hormones are deliberately injected into a hen, the hen will start behaving in male-like ways. The potentiality for behaving in a male-like manner must therefore have been present in the female; and by pressing (or, as students of behaviour usually say, 'releasing') the right button the environment can bring it out. But the higher parts of the brain respond to instructive stimuli: we *learn*.

Now let me carry the argument forward. It was a splendid idea to evolve into the possession of an organ that can respond to instructive stimuli, but the idea does not go far enough. If that were the whole story, we human beings might indeed live more successfully than other animals; but when we died, a new generation would have to start again from scratch. Let us go back for a moment to genetical instructions. A child at conception receives certain genetical instructions from its parents about how its growth and development are to proceed. Among these instructions there must be some which provide for the issue of further instructions; I mean, a child grows up in such a way that it, too, can eventually have children, and convey genetical instructions to them in turn. We are dealing here with a very special system of communication: a *hereditary* system. There are many examples of systems of this kind. A chain letter is perhaps the simplest: we receive a letter from a correspondent who asks us to write to a third party, asking him in turn to write a letter of the same kind to a fourth, and so on – a hereditary system. The most complicated example is provided by the human brain itself; for it does indeed act as intermediary in a hereditary system of its own. We do more than learn: we teach and hand on; tradition accumulates; we record information and wisdom in books.

Just as a hereditary system is a special kind of system of communcation – one in which the instructions provide for the issue of further instructions – so there is a specially important kind of hereditary system: one in which the instructions passed on from one individual to another change in some systematic way in the course of time. A hereditary system with this property may be said to be conducting or undergoing an *evolution*. Genetic systems of heredity often transact evolutionary changes; so also does the hereditary system that is mediated through the brain. I think it is most important to distinguish between four stages in the evolution of a brain. The nervous system began, perhaps, as an organ which responded only to elective stimuli from the environment; the animal that possessed it reacted instinctively or by rote, if at all. There then arose a brain which could begin to accept instructive stimuli from the outside world; the brain in this sense has dim and hesitant beginnings going far back in geological time. The third stage, entirely distinguishable, was the evolution of a non-genetical system of heredity, founded upon the fact that the most complicated brains can do more than merely receive instructions; in one way or another they make it possible for the instructions to be handed on. The existence of this system of heredity – of tradition, in its most general sense – is a defining characteristic of human beings, and it has been important for, perhaps, 500,000 years. In the fourth stage, not clearly distinguishable from the third, there came about a systematic change in the nature of the instructions passed on from generation to generation – an evolution, therefore, and one which has been going on at a great pace in the past 200 years. I shall borrow two words used for a slightly different purpose by the great demographer Alfred Lotka to distinguish between the two systems of heredity enjoyed by man: *endosomatic* or internal heredity for the ordinary or genetical heredity we have in common with other animals; and *exosomatic* or external heredity for the non-genetic heredity that is peculiarly our own – the heredity that is mediated through tradition, by which I mean the transfer of information through non-genetic channels from one generation to the next.

I am, of course, saying something utterly obvious: society changes; we pass on knowledge and skills and understanding from one person to another and from one generation to the next; a man can indeed influence posterity by other than genetic means. But I wanted to put the matter in a way which shows that we must not distinguish a strictly biological evolution from a social, cultural, or technological evolution: *both* are biological evolutions: the distinction between them is that the one is genetical and the other is not.

What, then, is to be inferred from all this? What lessons are to be learned from the similarities and correspondences between the two systems of biological heredity possessed by human beings? The answer is important, and I shall now try to justify it: the answer, I believe, is almost none.

It is true that a number of amusing (but in one respect highly dangerous) parallels can be drawn between our two forms of heredity and evolution. Just as biologists speak in a kind of shorthand about the 'evolution' of hearts or ears or legs – it is too clumsy and long-winded to say every time that these organs participate in evolution, or are outward expressions of the course of evolution – so we can speak of the evolution of bicycles or wireless sets or aircraft with the same qualification in mind: they do not really evolve, but they are appendages, exosomatic organs if you like, that evolve with us. And there are many correspondences between the two kinds of evolution. Both are gradual if we take the long view; but on closer inspection we shall find that novelties arise, not everywhere simultaneously – pneumatic tyres did not suddenly appear in the whole population of bicycles – but in a few members of the population: and if these novelties confer economic fitness, or fitness in some more ordinary and obvious sense, then the objects that possess them will spread through the population as a whole and become the prevailing types. In both styles of evolution we can witness an adaptive radiation, a deployment into different environments: there are wireless sets not only for the home, but for use in motor cars or for carrying about. Some great dynasties die out – airships, for example, in common with the dinosaurs they were so often likened to; others become fixed and stable: toothbrushes retained the same design and constitution for more than a hundred years. And, no matter what the cause of it, we can see in our exosomatic appendages something equivalent to vestigial organs; how else should we describe those functionless buttons on the cuffs of men's coats?

All this sounds harmless enough: why should I have called it dangerous? The danger is that by calling attention to the similarities, which are not profound, we may forget the *differences* between our two styles of heredity and evolution; and the differences between them are indeed profound. In their hunger for synthesis and systematization, the evolutionary philosophers of the nineteenth century[1] and some of their modern counterparts have missed the point: they thought that great lessons were to be learnt from similarities between Darwinian and social evolution; but it is from the differences that all the great lessons are to be learnt. For one thing, our newer style of evolution is Lamarckian in nature. The environment cannot imprint genetical information upon us, but it can and does imprint non-genetical information which we can and do pass on. Acquired characters are indeed inherited. The blacksmith was under an illusion if he supposed

---

[1] For example, Haeckel: 'The theory of selection teaches us that in human life, exactly as in animal and plant life, at each place and time only a small privileged minority can continue to exist and flourish; the great mass must starve and more or less prematurely perish in misery. . . . We may deeply mourn this tragic fact, but we cannot deny or alter it.'

that his habits of life could impress themselves upon the genetic make-up of his children; but there is no doubting his ability to teach his children his trade, so that they can grow up to be as stalwart and skilful as himself. It is because this newer evolution is so obviously Lamarckian in character that we are under psychological pressure to believe that genetical evolution must be so too. But although one or two biologists are still feebly trying to graft a Lamarckian or instructive interpretation upon ordinary genetical evolution, they are not nearly so foolish or dangerous as those who have attempted to graft a Darwinian or purely elective interpretation upon the newer, non-genetical, evolution of mankind.

The conception I have just outlined is, I think, a liberating conception. It means that we can jettison all reasoning based upon the idea that changes in society happen in the style and under the pressures of ordinary genetic evolution; abandon any idea that the direction of social change is governed by laws other than laws which have at some time been the subject of human decisions or acts of mind. That competition between one man and another is a necessary part of the texture of society; that societies are organisms which grow and must inevitably die; that division of labour within a society is akin to what we can see in colonies of insects; that the laws of genetics have an overriding authority; that social evolution has a direction imposed upon it by agencies beyond man's control – all these are biological judgments; but, I do assure you, bad judgments based upon a bad biology.

P. B. MEDAWAR, *The Future of Man*, 1960 (Reith Lectures, 1959)

CHAPTER NINE

Our Own Worst Enemies

1. The Perils of Progress

Science and the techniques to which it has given rise have changed human life during the last hundred and fifty years more than it had been changed since men took to agriculture, and the changes that are being wrought by science continue at an increasing speed. There is no sign of any new stability to be attained on some scientific plateau. On the contrary, there is every reason to think that the revolutionary possibilities of science extend immeasurably beyond what has so far been realized. Can the human race adjust itself quickly enough to these vertiginous transformations, or will it, as innumerable former species have done, perish from lack of adaptability? The dinosaurs were, in their day, the lords of creation, and if there had been philosophers among them not one would have foreseen that the whole race might perish. But they became extinct because they could not adapt themselves to a world without swamps. In the case of man and science, there is a wholly new factor, namely that man himself is creating the changes of environment to which he will have to adjust himself with unprecedented rapidity. But, although man through his scientific skill is the cause of the changes of environment, most of these changes are not willed by human beings. Although they come about through human agencies, they have, or at any rate have had so far, something of the inexorable inevitability of natural forces. Whether nature dried up the swamps or men deliberately drained them, makes little difference as regards the ultimate result. Whether men will be able to survive the changes of environment that their own skill has brought about is an open question. If the answer is in the affirmative, it will be known some day; if not, not. If the answer is to be in the affirmative, men will have to apply scientific ways of thinking to themselves and their institutions. They cannot continue to hope, as all politicians hitherto have, that in a world where everything has changed, the political and social habits of the eighteenth century can remain inviolate. Not only will men of science have to grapple with the sciences that deal with man, but – and this is a far more difficult matter – they will have to persuade the world to listen to what they have discovered. If they cannot succeed in this difficult enterprise, man will destroy himself by his halfway cleverness. I am told that, if he were out of the way, the future would lie with rats. I hope they will find it a pleasant world, but I am glad I shall not be there.

BERTRAND RUSSELL, *Science and Human Life* (in *What is Science*), 1956

The passages which follow outline two or three of the more obvious examples of what Bertrand Russell is talking about. This kind of physical threat (particularly, in the case of soil-erosion, when it is considered in conjunction with the present population explosion) is in all conscience alarming enough, but some of the more subtle and insidious dangers not here touched upon are perhaps even more so.

11. The Threat to the Soil

Erosion in Nature is a beneficient process without which the world would have died long ago. The same process, accelerated by human mismanagement, has become one of the most vicious and destructive forces that have ever been released by man. What is usually known as 'geological erosion' or 'denudation' is a universal phenomenon which through thousands of years has carved the earth into its present shape. Denudation is an early and important process in soil formation, whereby the original rock material is continuously broken down and sorted out by wind and water until it becomes suitable for colonization by plants. Plants, by the binding effects of their roots, by the protection they afford against rain and wind and by the fertility they impart to the soil, bring denudation almost to a standstill. Everybody must have compared the rugged and irregular shape of bare mountain peaks where denudation is still active with the smooth and harmonious curves of slopes that have long been protected by a mantle of vegetation. Nevertheless, some slight denudation is always occurring. As each superficial film of plant-covered soil becomes exhausted it is removed by rain or wind, to be deposited mainly in the rivers and sea, and a corresponding thin layer of new soil forms by slow weathering of the underlying rock. The earth is continuously discarding its old, worn-out skin and renewing its living sheath of soil from the dead rock beneath. In this way an equilibrium is reached between denudation and soil formation so that, unless the equilibrium is disturbed, a mature soil preserves a more or less constant depth and character indefinitely. The depth is sometimes only a few inches, occasionally several feet, but within it lies the whole capacity of the earth to produce life. Below that thin layer comprising the delicate organism known as soil is a planet as lifeless as the moon.

The equilibrium between denudation and soil formation is easily disturbed by the activities of man. Cultivation, deforestation or the destruction of natural vegation by grazing or other means, unless carried out according to certain immutable conditions imposed by each region, may so accelerate denudation that the soil, which would normally be washed or blown away in a century, disappears within a year or even within a day. But no human ingenuity can accelerate the soil-renewing process from lifeless rock to an extent at all comparable to the acceleration of denudation. This man-accelerated denudation is what is now known as soil erosion. It is the almost inevitable result of reducing below a certain limit the natural fertility of the soil – of man betraying his most sacred trust when he assumes dominion over the land. . . .

That the ultimate consequence of unchecked soil erosion, when it sweeps over whole countries as it is doing to-day, must be national extinction is obvious, for whatever other essential raw material a nation may dispense with, it cannot exist without fertile soil. Nor is extinction of a nation by erosion merely a hypothetical occurrence that may occur at some future date; it has occurred several times in the past. Erosion has, indeed, been one of the most potent factors causing the downfall of former civilizations and empires whose ruined cities now lie amid barren wastes that once were the world's most fertile lands. The deserts of North China, Persia, Mesopotamia, and North Africa tell the same story of the gradual exhaustion of the soil as the increasing demands made upon it by expanding civilization exceeded its recuperative powers. Soil erosion, then as now, followed soil exhaustion. . . .

For erosion is the modern symptom of maladjustment between human society and its environment. It is a warning that Nature is in full revolt against the sudden incursion of an exotic civilization into her ordered domains. Men are permitted to dominate Nature on precisely the same condition as trees and plants, namely on condition that they improve the soil and leave it a little better for their posterity than they found it. Agriculture in Europe, whatever its other weaknesses, has been, and perhaps still is, a practice tending on the whole to increase soil fertility.[1] When adopted and adapted elsewhere it has resulted, almost invariably, in a catastrophic decrease in fertility. The illusion that fertility can always be restored by applying some of the huge amounts of artificial fertilizers now available has been shattered by the recognition that fertility is not merely a matter of plant-food supply (for even exhausted soils usually contain ample reserves of plant food), but is also closely connected with soil *stability*. An exhausted soil is an unstable soil; Nature has no further use for it and removes it bodily. The process is the same as denudation, but whereas under normal conditions a fraction of an inch of soil may become exhausted and be removed in a century, under human control the entire depth of soil may become exhausted and be eroded in a few years.

G. V. JACKS and R. O. WHYTE, *The Rape of the Earth*, 1939

---

[1] *As the authors point out elsewhere, this is due less to altruism and farsightedness on the part of European farmers than to accidentally favourable combinations of factors such as a temperate climate (extremes erode so much more quickly) and forest soils (requiring hard work – unsuitable for intensive exploitation and quick returns). (Ed.)*

III. The Threat to the Race

The history of life on earth has been a history of interaction between living things and their surroundings. To a large extent, the physical form and the habits of the earth's vegetation and its animal life have been moulded by the environment. Considering the whole span of earthly time, the opposite effect, in which life actually modifies its surroundings, has been relatively slight. Only within the moment of time represented by the present century has one species – man – acquired significant power to alter the nature of his world.

During the past quarter-century this power has not only increased to one of disturbing magnitude but it has changed in character. The most alarming of all man's assaults upon the environment is the contamination of air, earth, rivers, and sea with dangerous and even lethal materials. This pollution is for the most part irrecoverable; the chain of evil it initiates not only in the world that must support life but in living tissues is for the most part irreversible. In this now universal contamination of the environment, chemicals are the sinister and little-recognized partners of radiation in changing the very nature of the world – the very nature of its life. Strontium 90, released through nuclear explosions into the air, comes to earth in rain or drifts down as fallout, lodges in soil, enters into the grass or corn or wheat grown there, and in time takes up its abode in the bones of a human being, there to remain until his death. Similarly, chemicals sprayed on croplands or forests or gardens lie long in soil, entering into living organisms, passing from one to another in a chain of poisoning and death. Or they pass mysteriously by underground streams until they emerge and, through the alchemy of air and sunlight, combine into new forms that kill vegetation, sicken cattle, and work unknown harm on those who drink from once-pure wells. As Albert Schweitzer has said, 'Man can hardly recognize the devils of his own creation.'

It took hundreds of millions of years to produce the life that now inhabits the earth – aeons of time in which that developing and evolving and diversifying life reached a state of adjustment and balance with its surroundings. The environment, rigorously shaping and directing the life it supported, contained elements that were hostile as well as supporting. Certain rocks gave out dangerous radiation; even within the light of the sun, from which all life draws its energy, there were short-wave radiations with power to injure. Given time – time not in years but in millennia – life adjusts, and a balance has been reached. For time is the essential ingredient; but in the modern world there is no time.

The rapidity of change and the speed with which new situations are created follow the impetuous and heedless pace of man rather than the deliberate pace of nature. Radiation is no longer merely the background radiation of rocks, the bombardment of cosmic rays, the ultra-violet of the sun that have existed before there was any life on earth; radiation is now the unnatural creation of man's tampering with the atom. The chemicals to which life is asked to make its adjustment are no longer merely the calcium and silica and copper and all the rest of the minerals washed out of the rocks and carried in rivers to the sea; they are the synthetic creations of man's inventive mind, brewed in his laboratories, and having no counterparts in nature.

To adjust to these chemicals would require time on the scale that is nature's; it would require not merely the years of a man's life but the life of generations. And even this, were it by some miracle possible, would be futile, for the new chemicals come from our laboratories in an endless stream; almost five hundred annually find their way into actual use in the United States alone. The figure is staggering and its implications are not easily grasped – five hundred new chemicals to which the bodies of men and animals are required somehow to adapt each year, chemicals totally outside the limits of biological experience.

Among them are many that are used in man's war against nature. Since the mid-1940s over two hundred basic chemicals have been created for use in killing insects, weeds, rodents, and other organisms described in the modern vernacular as 'pests'; and they are sold under several thousand different brand names.

These sprays, dusts, and aerosols are now applied almost universally to farms, gardens, forests, and homes – non-selective chemicals that have the power to kill every insect, the 'good' and the 'bad', to still the song of birds and the leaping of fish in the streams, to coat the leaves with a deadly film and to linger on in soil – all this though the intended target may be only a few weeds or insects. Can anyone believe it is possible to lay down such a barrage of poisons on the surface of the earth without making it unfit for all life? They should not be called 'insecticides', but 'biocides'.

The whole process of spraying seems caught up in an endless spiral. Since DDT was released for civilian use, a process of escalation has been going on in which ever more toxic materials must be found. This has happened because insects, in a triumphant vindication of Darwin's principle of the survival of the fittest, have evolved super races immune to the particular insecticide used, hence a deadlier one has always to be developed – and then a deadlier one than that. It has happened also because, for reasons to be described later,[1] destructive insects often undergo a 'flareback' or resurgence, after spraying, in numbers greater than before. Thus the chemical war is never won, and all life is caught in its violent crossfire.

Along with the possibility of the extinction of mankind by nuclear war, the central problem of our age has therefore become the contamination of man's total environment with such substances of incredible potential for harm – substances that accumulate in the tissues of plants and animals and even penetrate the germ cells to shatter or alter the very material of heredity upon which the shape of the future depends.

Some would-be architects of our future look towards a time when it will be possible to alter the human germ plasm by design. But we may easily be doing so now by inadvertence, for many chemicals, like radiation, bring about gene mutations. It is ironic

---

[1] *The spray used against an enemy of plant life may prove even more effective in eliminating an enemy of the enemy.* (*Ed.*)

to think that man might determine his own future by something
so seemingly trivial as the choice of an insect spray.

RACHEL CARSON, *Silent Spring*, 1962

The Rape of the Earth *and* Silent Spring : *the very titles seem
to give fair warning of the authors' designs on us. Is this impression
borne out by the styles employed? How do these compare with
Russell's effortless clarity and octogenarian detachment?*

CHAPTER TEN
Science and Art

Both the passages in this chapter deal with aspects of the clash, real or imagined, between science and the arts or humanities, though the aspects are different, and there has been no attempt to choose two extracts in direct opposition to one another. Both the authors represented have a foot in either camp, though in each case one seems rather more firmly planted than the other.

There is an interesting contrast in styles, the one employing a deliberately flat, throw-away technique, the other having a much more richly patterned texture.

1. The Two Cultures

Literary intellectuals at one pole – at the other scientists, and as the most representative, the physical scientists. Between the two a gulf of mutual incomprehension – sometimes (particularly among the young) hostility and dislike, but most of all lack of understanding. . . .

The non-scientists have a rooted impression that the scientists are shallowly optimistic, unaware of man's condition. On the other hand, the scientists believe that the literary intellectuals are totally lacking in foresight, peculiarly unconcerned with their brother men, in a deep sense anti-intellectual, anxious to restrict both art and thought to the existential moment. And so on . . .

First, about the scientists' optimism. This is an accusation which has been made so often that it has become a platitude. It has been made by some of the acutest non-scientific minds of the day. But it depends upon a confusion between the individual experience and the social experience, between the individual condition of man and his social condition. Most of the scientists I have known well have felt – just as deeply as the non-scientists I have known well – that the individual condition of each of us is tragic. Each of us is alone: sometimes we escape from solitariness, through love or affection or perhaps creative moments, but those triumphs of life are pools of light we make for ourselves while the edge of the road is black: each of us dies alone. Some scientists I have known have had faith in revealed religion. Perhaps with them the sense of the tragic condition is not so strong. I don't know. With most people of deep feeling, however high-spirited and happy they are, sometimes most with those who are happiest and most high-spirited, it seems to be right in the fibres, part of the weight of life. That is as true of the scientists I have known best as of anyone at all.

But nearly all of them – and this is where the colour of hope genuinely comes in – would see no reason why, just because the individual condition is tragic, so must the social condition be. Each of us is solitary: each of us dies alone: all right, that's a fate against which we can't struggle – but there is plenty in our condition which is not fate, and against which we are less than human unless we do struggle.

Most of our fellow human beings, for instance, are under-fed and die before their time. In the crudest terms, *that* is the social condition. There is a moral trap which comes through the insight into a man's loneliness: it tempts one to sit back, complacent in one's unique tragedy, and let the others go without a meal.

As a group, the scientists fall into that trap less than others. They are inclined to be impatient to see if something can be done: and inclined to think that it can be done, until it's proved otherwise. That is their real optimism, and it's an optimism that the rest of us badly need. . . .

. . . The scientific culture really is a culture, not only in an intellectual but also in an anthropological sense. That is, its members need not, and of course often do not, always completely understand each other; biologists more often than not will have a pretty hazy idea of contemporary physics; but there are common attitudes, common standards and patterns of behaviour, common approaches and assumptions. This goes surprisingly wide and deep. It cuts across other mental patterns, such as those of religion or politics or class.

Statistically, I suppose slightly more scientists are in religious terms unbelievers, compared with the rest of the intellectual world – though there are plenty who are religious, and that seems to be increasingly so among the young. Statistically also, slightly more scientists are on the Left in open politics – though again, plenty always have called themselves conservatives, and that also seems to be more common among the young. Compared with the rest of the intellectual world, considerably more scientists in this country and probably in the U.S. come from poor families. Yet, over a whole range of thought and behaviour, none of that matters very much. In their working, and in much of their emotional life, their attitudes are closer to other scientists than to non-scientists who in religion or politics or class have the same labels as themselves. If I were to risk a piece of shorthand, I should say that naturally they had the future in their bones.

They may or may not like it, but they have it. That was as true of the conservatives J. J. Thomson and Lindemann as of the radicals Einstein or Blackett: as true of the Christian A. H. Compton as of the materialist Bernal: of the aristocrats Broglie or Russell as of the proletarian Faraday: of those born rich, like Thomas Merton or Victor Rothschild, as of Rutherford, who was the son of an odd-job handyman. Without thinking about it, they respond alike. That is what a culture means.

At the other pole, the spread of attitudes is wider. It is obvious that between the two, as one moves through intellectual society from the physicist to the literary intellectuals, there are all kinds of tones of feeling on the way. But I believe the pole of total incomprehension of science radiates its influence on all the rest. That total incomprehension gives, much more pervasively than we realise, living in it, an unscientific flavour to the whole 'traditional' culture, and that unscientific flavour is often, much more than we admit, on the point of turning anti-scientific. The feelings of one pole become the anti-feelings of the other. If the scientists have the future in their bones, then the traditional culture responds by wishing the future did not exist.[1] It is the

[1] Compare George Orwell's *1984*, which is the strongest possible wish that the future should not exist, with J. D. Bernal's *World Without War*.

traditional culture, to an extent remarkably little diminished by the emergence of the scientific one, which manages the western world. . . .

The degree of incomprehension on both sides is the kind of joke which has gone sour. There are about fifty thousand working scientists in the country and about eighty thousand professional engineers or applied scientists. During the war and in the years since, my colleagues and I have had to interview somewhere between thirty to forty thousand of these – that is, about 25 per cent. The number is large enough to give us a fair sample, though of the men we talked to most would still be under forty. We were able to find out a certain amount of what they read and thought about. I confess that even I, who am fond of them and respect them, was a bit shaken. We hadn't quite expected that the links with the traditional culture should be so tenuous, nothing more than a formal touch of the cap.

As one would expect, some of the very best scientists had and have plenty of energy and interest to spare, and we came across several who had read everything that literary people talk about. But that's very rare. Most of the rest, when one tried to probe for what books they had read, would modestly confess, 'Well, I've *tried* a bit of Dickens', rather as though Dickens were an extra-ordinarily esoteric, tangled and dubiously rewarding writer, something like Rainer Maria Rilke. In fact that is exactly how they do regard him: we thought that discovery, that Dickens had been transformed into the type-specimen of literary incomprehensi-bility, was one of the oddest results of the whole exercise.

But of course, in reading him, in reading almost any writer whom we should value, they are just touching their caps to the traditional culture. They have their own culture, intensive, rigorous, and constantly in action. This culture contains a great deal of argument, usually much more rigorous, and almost always at a higher conceptual level, than literary persons' argu-ments – even though the scientists do cheerfully use words in senses which literary persons don't recognise, the senses are exact ones, and when they talk about 'subjective', 'objective', 'philosophy' or 'progressive',[1] they know what they mean, even though it isn't what one is accustomed to expect.

Remember, these are very intelligent men. Their culture is in many ways an exacting and admirable one. It doesn't contain much art, with the exception, an important exception, of music. Verbal exchange, insistent argument. Long-playing records. Colour-photography. The ear, to some extent the eye. Books, very little, though perhaps not many would go so far as one hero, who perhaps I should admit was further down the scientific ladder than the people I've been talking about – who, when asked what books he read, replied firmly and confidently: 'Books? I prefer to use my books as tools.' It was very hard not to let the mind wander – what sort of tool would a book make? Perhaps a hammer? A primitive digging instrument?

Of books, though, very little. And of the books which to most literary persons are bread and butter, novels, history, poetry, plays, almost nothing at all. It isn't that they're not interested in the psychological or moral or social life. In the social life, they certainly are, more than most of us. In the moral, they are by and large the soundest group of intellectuals we have; there is a moral component right in the grain of science itself, and almost all scientists form their own judgments of the moral life. In the psychological they have as much interest as most of us, though occasionally I fancy they come to it rather late. It isn't that they lack the interests. It is much more that the whole literature of the traditional culture doesn't seem to them relevant to those interests. They are, of course, dead wrong. As a result, their imaginative understanding is less than it could be. They are self-impoverished.

But what about the other side? They are improverished too – perhaps more seriously, because they are vainer about it. They still

---

[1] *Subjective*, in contemporary technological jargon, means 'divided according to subjects'. *Objective* means 'directed towards an object'. *Philosophy* means 'general intellectual approach or attitude' (for example, a scientist's 'philosophy of guided weapons' might lead him to propose certain kinds of 'objective research'.) A 'progressive' job means one with possibilities of promotion.

like to pretend that the traditional culture is the whole of 'culture', as though the natural order didn't exist. As though the exploration of the natural order was of no interest either in its own value or its consequences. As though the scientific edifice of the physical world was not, in its intellectual depth, complexity and articulation, the most beautiful and wonderful collective work of the mind of man. Yet most non-scientists have no conception of that edifice at all. Even if they want to have it, they can't. It is rather as though, over an immense range of intellectual experience, a whole group was tone-deaf. Except that this tone-deafness doesn't come by nature, but by training, or rather the absence of training.

As with the tone-deaf, they don't know what they miss. They give a pitying chuckle at the news of scientists who have never read a major work of English literature. They dismiss them as ignorant specialists. Yet their own ignorance and their own specialisation is just as startling. A good many times I have been present at gatherings of people who, by the standards of the traditional culture, are thought highly educated and who have with considerable gusto been expressing their incredulity at the illiteracy of scientists. Once or twice I have been provoked and have asked the company how many of them could describe the Second Law of Thermodynamics. The response was cold: it was also negative. Yet I was asking something which is about the scientific equivalent of: *Have you read a work of Shakespeare's?*

I now believe that if I had asked an even simpler question – such as, What do you mean by mass, or acceleration, which is the scientific equivalent of saying, *Can you read?* – not more than one in ten of the highly educated would have felt that I was speaking the same language. So the great edifice of modern physics goes up, and the majority of the cleverest people in the western world have about as much insight into it as their neolithic ancestors would have had.

C. P. SNOW, *The Two Cultures and the Scientific Revolution*, 1959 (Rede Lecture, 1959)

11. The Two Brains

(*a*) The brain's anatomy closely reproduces its own past history: that is its greatest fascination. One can say: 'So does the entire body, there is no part of us which is not the creation of time.' True, yet the historical nature of the structure within the skull is more dramatic, for here the ancient parts survive in almost their original form and the new have been added to them, as in a Gothic cathedral with a Norman nave, Decorated choir, and Perpendicular aisles and transepts. Thus with peculiar vividness it shows how life defeats time, its present moment always embracing the past's whole achievement. The embryo repeats in rough summary the events of hundreds of millions of years in nine months, but the brain maintains them all simultaneously. Our past life from worms through fishes and amphibians towards mammals is summed up in the Old Brain, while the advance towards full humanity is expressed in the elaboration of the New Brain, these convoluted hemispheres that dwarf and all but conceal their primitive antecedents.

In an earlier chapter I have described how the spinal cord of a creature resembling the little mud-lurking lancelet was the beginning of the Central Nervous System and how, as that system acquired more duties in controlling the bodily movement and servicing the senses, the upper end of the spinal cord developed the nobby ganglion that was the beginning of brain. By the time of the fishes this had become a true if elementary brain, duly housed in a protective skull; among the amphibians with four jointed limbs and a head to be moved and rapidly sharpening terrestrial senses, it extended further forward again within a skull now massive and sharply defined. It is to be expected, then, that in man the portion of the brain lying where the spinal cord leaves the flexible tunnel of the vertebrae and passes through the neat hole left open for it in the floor of the skull should be the most primitive. Here are the Medulla and Pons, survivals from the worm stage of our ancestry, while the Cerebellum, hanging a little below, was present already in the fishy phase. Above the Pons lie the four nobs of the Corpora Quadrigemina, developed by the amphibians, and beyond and below them the Thalamus, the last and largest department of the Old Brain that underlies the New Brain and was also present already in the amphibians. . . . Growing from the Thalamus but extending over the whole Old Brain like some crinkled mob cap is the New Brain, the Cerebrum or Cerebral Cortex, which, from tiny beginnings in the more evolved amphibians, was enlarged in the mammals and then in the primates, but has attained its overwhelming size and import-ance only in man. . . .

Day and night the most ancient part of the brain, the Medulla and Pons, is at work incognito regulating the affairs of the body, the working of its myriad constituents. So long as it is intact the lungs will breathe and the heart beat even though the conscious mind is out of action from sleep, shock, or anaesthetics. The Cerebellum, working in conjunction with the coiled tubes of the semicircular Canals set inside the ear, is principally concerned with the balance and posture of the body and its movements through space. In man it is relatively small, showing like a little chignon below the mass of the Cerebrum, but in birds, as one might suppose, it is a large and highly important part of the brain. The Corpora Quadrigemina, once the seat of such consciousness as there was in the world, serves an obscure purpose in the human brain, being directly responsible only for such minor duties as the contraction of the pupils of the eye.

The Thalamus and Hypothalamus, although part of the Old Brain, play a role of the greatest importance in our conscious lives. They serve as a kind of intermediary between the unconscious life of the body and the Cerebrum, the prime seat of consciousness. Little known in the world, hidden out of sight below the envelop-ing folds of the cerebrum, the Thalamus has yet been recognized as 'the power behind the throne of human nature'. It is above all concerned with feeling, with the great, all-pervading emotions of anger, fear, and pleasure and so with giving the individual his sense of the tone and colour of life. While all the lower brain centres are standardized and vary little between man and man, the Thalamus is already highly individual, and indeed appears to inspire all the deeper levels of personality, those mysterious qualities we may call weight or density, which are so universally apprehended yet which remain so difficult to define in words. The thalamic brain is feminine in its nature, in contrast with the essential masculinity of the Cerebrum, or, in the old phrase, it represents heart as opposed to head. Here, in this dark, 'hidden chamber' surely lies the belly with which D. H. Lawrence would have us think. A Cerebrum so overdeveloped as to inhibit the Thalamus underlies that brittleness, silliness, and lack of true feeling characteristic of the pure 'intellectual', whom we expect to be as lacking in wisdom as he is genuinely clever. . . .

From the majestic seat of feeling and personality in the Old Brain we rise now to where intellect is enthroned in the New. While the Old Brain in its entirety weighs only some 175 grams in a brain of average size, the twin lobes of the Cerebrum or Cortex weigh as much as 1,200 grams. In the most cerebral of animals, the apes, the New Brain reaches about 350 grams; by such a head do we outstrip our fellow creatures.

(*b*) Any intelligent being from another planet would, I think, be puzzled by two quite opposite ideas to be found in our attempts to reconstruct our own history. Some historians write as though the story of man on earth was a fairly continuous progress, with invention added to invention, idea to idea and, as a result, an ever mounting control and understanding of nature. This has its truth. We know how we have gone from stone through bronze to iron and steel in the stuff of our tools; we know that after the invention of the bow and arrow in the Old Stone Age and weaving and potting in the New, our ancestors added the plough, the wheel, the sail, the working of copper and bronze during the fourth and third millennia B.C.; we know they went from magic and tribal daimons to the worship of high gods, from empirical mathematics and astronomy to all the fine complexities of theoretical science. It is extraordinary, indeed, how tenacious we have been of our skills and knowledge: in spite of our violent and bloody history, in spite of the many times when barbarians have put the civilized, the learned, and skilful to the sword, or neighbouring peoples ruined one another by ceaseless and idiotic warfare, few of the inventions made by men of genius have ever disappeared altogether from the earth; even knowledge and intellectual methods, though they have fluctuated from age to age, have seldom been quite extinguished from the human brain. Somewhere some few individuals kept them alive even through the darkest times. Thus, not only have there been wheeled vehicles on the roads, sailing-ships on the waters for the past five thousand years, but our mathematicians, physicists, astronomers, and surgeons still use knowledge which has been in circulation since it was painfully worked out by the ancient priesthoods.

All these are facts, and may seem to contradict what I have said about the fragility of culture. Yet other historians are able to write our story as though it were composed of a series of disjointed episodes, the perpetual rise and fall of peoples and cultures. And they, too, speak some part of the truth. There is no need to accept Spengler's automatic phases of rise and eclipse, nor Toynbee's twenty-one cultures with their peculiar mechanisms, to be aware of recurrent loss in history as well as continuous gain. Indeed all of us with any wisdom know that in many of the highest and most precious of human gifts and attributes we today are no more advanced, but indeed often much poorer, than peoples who held the flame of consciousness hundreds or thousands of years before us.

My friend Gordon Childe ended one of his most remarkable books with these sentences. 'Progress is real if discontinuous. The upward curve resolves itself into a series of troughs and crests. But in those domains that archaeology as well as written history can survey, no trough ever declines to the low level of the preceding one; each crest out-tops its last precursor.' They were written before we entered into the shadow of the atom bomb, so I am not blaming the author for over-optimism or failing to allow for the last wave rising to destroy the planet, but I think these sentences do confuse two quite distinct elements in our human faculties: the imaginative and the intellectual.

Professor Childe and those who think like him attach the highest importance to purely intellectual achievements and to the skill and techniques which they control. Here the gain has been continuous. Any reasonably capable student now at college could put the finest Greek mathematician to shame, any schoolchild could outdo the wise men of Sumer and Egypt. It is in the things of the imagination, of the complete psyche that the spirit bloweth where it listeth. No living artist would dare to claim equality with Mozart, Rembrandt, or Aeschylus; it would be a presumptuous sculptor who was confident of surpassing the masters of the Egyptian Old Kingdom. Great and small, imaginative powers may ebb away, and no effort of will, it seems, can regain them. . . .

Looking again inside the skull at the immediate source of all our abilities and disabilities, it seems evident that while the purely intellectual output of the Cerebrum, which can be exactly stated in words and recorded in writing, is usually maintained and tends to be heightened and refined from generation to generation, creations in which the Old Brain with its emotions and bodily relationship, its power over the unconscious are involved, creations, that is to say of high imagination or sensibility, appear often to wane with the old age of a people as they so commonly do with the old age of an individual artist. And, while one honours the intellectual achievement of our species, it is the works of imagination and feeling that give cultures their distinctive flavour, colour, and form, their power to delight and inspire. It is through them that a people is mainly remembered and judged. In their pure imaginative essence there is no progress in these things; from the age of the cave-paintings to the age of the Impressionists, from the Congo to China, genius has merely flowered and flowered again with all the riotous variety of the garden, while intellectual attainment has gone forward – erratically, but forward.

JACQUETTA HAWKES, *Man on Earth*, 1954

CHAPTER ELEVEN
Science and Religion

Again there has been no attempt to choose passages which will carry on a debate with one another. Antagonists over religion refuse to talk the same terminology or come to real grips with one another even when face to face, let alone when writing different books at different times in different languages. But at least Freud and Coulson take up standpoints or attitudes to religion which are openly and diametrically opposed to one another, even while dealing with quite different aspects to the controversy.

Whether Harrison succeeds (indeed, whether any man can succeed) in striking a real balance, or whether in being all things to all men he cannot help but be no thing to any man, is open to debate.

1. God the Father-Figure

If one wishes to form a true estimate of the full grandeur of religion, one must keep in mind what it undertakes to do for men. It gives them information about the source and origin of the universe, it assures them of protection and final happiness amid the changing vicissitudes of life, and it guides their thoughts and actions by means of precepts which are backed by the whole force of its authority. It fulfils, therefore, three functions. In the first place, it satisfies man's desire for knowledge; it is here doing the same thing that science attempts to accomplish by its own methods, and here, therefore, enters into rivalry with it. It is to the second function that it performs that religion no doubt owes the greater part of its influence. In so far as religion brushes away men's fear of the dangers and vicissitudes of life, in so far as it assures them of a happy ending, and comforts them in their misfortunes, science cannot compete with it. Science, it is true, teaches how one can avoid certain dangers and how one can combat many sufferings with success; it would be quite untrue to deny that science is a powerful aid to human beings, but in many cases it has to leave them to their suffering, and can only advise them to submit to the inevitable. In the performance of its third function, the provision of precepts, prohibitions and restrictions, religion is furthest removed from science. For science is content with discovering and stating the facts. It is true that from the applications of science rules and recommendations for behaviour may be deduced. In certain circumstances they may be the same as those which are laid down by religion, but even so the reasons for them will be different.

It is not quite clear why religion should combine these three functions. What has the explanation of the origin of the universe to do with the inculcation of certain ethical precepts? Its assurances of protection and happiness are more closely connected with these precepts. They are the reward for the fulfilment of the commands; only he who obeys them can count on receiving these benefits, while punishment awaits the disobedient. For the matter of that something of the same kind applies to science; for it declares that anyone who disregards its inferences is liable to suffer for it.

One can only understand this remarkable combination of teaching, consolation and precept in religion if one subjects it to genetic analysis. We may begin with the most remarkable item of the three, the teaching about the origin of the universe – for why should a cosmogony be a regular element of religious systems? The doctrine is that the universe was created by a being similar to man, but greater in every respect, in power, wisdom and strength of passion, in fact, by an idealized superman. Where you have animals as creators of the universe, you have indications of the influence of totemism, which I shall touch on later, at any rate with a brief remark. It is interesting to note that this creator is nearly always a male, although there is no lack of indication of the existence of female deities, and many mythologies make the creation of the world begin precisely with a male god triumphing over a female goddess, who is degraded into a monster. This raises the most fascinating minor problems, but we must hurry on. The rest of our enquiry is made easy because this God-Creator is openly called Father. Psycho-analysis concludes that he really is the father, clothed in the grandeur in which he once appeared to the small child. The religious man's picture of the creation of the universe is the same as his picture of his own creation.

If this is so, then it is easy to undertstand how it is that the comforting promises of protection and the severe ethical commands are found together with the cosmogony. For the same individual to whom the child owes its own existence, the father (or, more correctly, the parental function which is composed of the father and the mother), has protected and watched over the weak and helpless child, exposed as it is to all the dangers which threaten in the external world; in its father's care it has felt itself safe. Even the grown man, though he may know that he possesses greater strength, and though he has greater insight into the dangers of life, rightly feels that fundamentally he is just as helpless and unprotected as he was in childhood and that in relation to the external world he is still a child. But he has long ago realized that his father is a being with strictly limited powers and by no means endowed with every desirable attribute. He therefore looks back to the memory-image of the overrated father of his childhood, exalts it into a Deity, and brings it into the present and into reality. The emotional strength of this memory-image and the lasting nature of his need for protection are the two supports of his belief in God.

The third main point of the religious programme, its ethical precepts, can also be related without any difficulty to the situation of childhood. In a famous passage, which I have already quoted in an earlier lecture, the philosopher Kant speaks of the starry heaven above us and the moral law within us as the strongest evidence for the greatness of God. However odd it may sound to put these two side by side – for what can the heavenly bodies have to do with the question whether one man loves another or kills him? – nevertheless it touches on a great psychological truth. The same father (parental function) who gave the child his life, and preserved it from the dangers which that life involves, also taught it what it may or may not do, made it accept certain limitations of its instinctual wishes, and told it what consideration it would be expected to show towards its parents and brothers and sisters, if it wanted to be tolerated and liked as a member of the family circle, and later on of more extensive groups. The child is brought up to know its social duties by means of a system of love-rewards and punishments, and in this way it is taught that its security in life depends on its parents (and, subsequently, other people) loving it and being able to believe in its love for them. This whole state of affairs is carried over by the grown man unaltered into his religion. The prohibitions and commands of his parents live on in his breast as his moral conscience; God rules the world of men with the help of the same system of rewards and punishments, and the degree of protection and happiness which each individual enjoys depends on his fulfilment of the demands of morality; the feeling of security, with which he fortifies himself against the dangers both of the external world and of his human environment, is founded on his love of God and the consciousness of God's love for him. Finally, he has in prayer a direct influence on the divine will, and in that way insures for himself a share in the divine omnipotence.

SIGMUND FREUD, *New Introductory Lectures on Psycho-Analysis*, 1933 (tr. W. J. H. Sprott)

11. The Lamp of the World

Most dogmas represent useful holding operations designed to help suffering humanity visualize the eternal truths whose contemplation can give solace to the human spirit. A dogma is the emotional equivalent of a shelter, erected to protect a certain area of thought space in which are placed precious and needed utensils, and to hold away the great dark void of the unknown that presses about us all. Hence dogmas have a very important place in the erection and perpetuation of a religion that is to minister to the needs of the common man.

But dogmas retained too long restrain growth, or cause pain where growth occurs. The shell of a crustacean, useful to protect its soft body, becomes restrictive as it grows. This plastic armour must be split if expansion is to be accommodated. Science provides useful mechanisms for splitting outmoded dogmas. All of us need to shed our spiritual shells at times and grow larger ones, but we are likely when we do this to feel considerable tenderness until the new shell hardens.

Constant vigilance is needed on the part of priest and preacher to keep the external expression of the inner principles of religion in pace with man's development. Science helps in this by constantly feeding new information about man and nature into religion, and producing growth which prevents dogmas from remaining crystallized too long. Understandably some religious leaders resist such change, for they are mortal custodians of eternal verities, and can be expected to resent the cracking of the literalness that stiffens the embodiment of spiritual truth.

We may ask the scientist who resents the worshiping of what seem to him false images of real truths, 'What choice has man, with his limited comprehension as he toils upward on his path, but to get his light from the lamps he sees about him? And who can blame him if during the earlier parts of his climb he confuses the lamp with the light?'

GEORGE RUSSELL HARRISON, *What Man May Be*, 1957

III. A Question of Perspective

A few years ago I was partly responsible for the construction of an underground laboratory. It was at King's College, London University, and its rather curious location, directly underneath the main College quadrangle, was forced upon us by the exigencies of space in central London. While the laboratory was in course of construction, we had frequent occasion to consult the architect, and look through the large sheaf of drawings that he had in his office. Some of these were plans, showing us what the floor space would look like to an imaginary observer overhead: others were elevations, from one side or one end; or they were sections, in different directions and at different levels. Many of the diagrams looked utterly unlike the others: some showed features not present in the rest. Occasionally there were substantial common elements as when a plan and an elevation showed the existence of a boundary wall. Some drawings, from their very nature, showed a lot of detail; others showed relatively little; but, so far as the architect could make them, each was complete. None of them was exhaustive, and it would always be possible to imagine additional drawings, as for example sections in a different direction, which would resemble existing drawings in greater or less degree, though they would not be identical with any.

Now despite all these differences, we know perfectly well that there is only one building. These are representations of it, in the form appropriate to a piece of paper which is only two-dimensional. We need to have several of these drawings before we can say that we know what the building is really like. From one point of view, not a single one among all these drawings is ultimately redundant; and every drawing will have something particular to tell us about the building. To the uninitiated, it will seem almost impossible that all of these several descriptions can be 'true', though in fact they are; or that they represent one building, as in fact they do. . . .

It may be worth giving a simple example to show the complementary character of the various accounts that can be given of one situation. I choose a primrose, because this will enable me to bring out the widely differing characteristics of what are, in a sense, parallel interpretations. To the question, 'what is a primrose?' several valid answers may be given. One person says:

> A primrose by the river's brim
> A yellow primrose was to him,
> And it was nothing more.

Just that, and no more. Another person, the scientist, says 'a primrose is a delicately balanced biochemical mechanism, requiring potash, phosphates, nitrogen and water in definite proportions.' A third person says 'a primrose is God's promise of spring.' All three descriptions are correct, but they have about as much in common as three quite separate sections of the underground physics laboratory.

When they have thought about their work, many of the best scientists have recognised this alternative character of the descriptions which they give. Among the physicists it has become an almost universally accepted item of belief. For example we can point to the great controversy about light – was it corpuscular, as Newton believed; or wave-like, as Huygens claimed? Certainly some phenomena were better understood in one language, other phenomena in the alternative. But now a dualism is accepted: we use either the one or the other, choosing that which is best adapted to our particular situation. This does not mean that light is *both* corpuscle and wave: the dualism lies not in what Kant would call 'the-thing-in-itself', but in our interpretation of it, in the language and concepts that we use to give meaning and pattern to experiments in optics and spectroscopy. It is just the same with the celebrated discussion about the nature of an electron; is that also a particle or is it a wave? The answer is precisely the same as before. 'We don't know.' For the thing-in-itself is as much unknown to us as is the physics laboratory at King's College, London, which, being entirely underground, cannot be 'seen' in the usual sense of that word. What physicists have done is to devise models whereby the behaviour of the electron may be fitted into a pattern: and they have found that two different sets of concepts are needed to do justice to this behaviour. There was a time when this duality of description would have been rejected as wholly improper: and even now an occasional voice, such as that of Einstein, is raised against one or other aspect of the duality. But most of us have lived so long with this that we have grown used to it; and have come to see the great and liberating influence inherent in the two modes of description. . . .

Perhaps it is natural that the physicist should have been the first to remind us that, even within science – even within one branch of science – this concept of 'sections of a building' must be introduced. For physics was the earliest science to develop any thorough-going discipline. But other scientists are coming to see it now, in greater numbers than before. Here, for example, is Tinbergen writing an account of what we mean by Instinct.[1] He begins by distinguishing three ways of studying behaviour that may be called instinctive. First we may seek the 'causal structure' underlying it; what we might loosely refer to as the physics and chemistry of instinct. Then we may concentrate on the directiveness or biological purposiveness associated with it. This, which has a teleological character, is most important in any complete study of behaviour, but it is not a substitute for causal study. Finally there is the psychological way, quite distinct from the other two. . . .

I believe that when Niels Bohr introduced this idea of complementarity into physics, and then extended it to apply more widely, he was opening a new chapter in our understanding of the universe we live in. Many of the celebrated debates of former days, and the struggles of to-day, are essentially examples of this duality. Both sides are right, but they have no real contact with each other; and their points of view are like two distinct sections of our physics laboratory. . . . The whole matter is so important that I would like to illustrate it by reference to three such celebrated debates – mind and matter, free-will and determinism, and teleology.

First let us consider mind and matter. The issue is perfectly simple: mind is associated with body and brain, for we have no direct physical experience of a disembodied mind. But the brain is a most intricate collection of some 10^{11} of tiny electrical circuits, composed of nerves and so ultimately of atoms and molecules. The nature of a thought can be said to correspond to the patterns of electrical currents in these many circuits: for example, the activity of millions of neurons is involved in the recall of any memory; and sanity and insanity can be distinguished by the electro encephalograph and revealed in the different kinds of rhythm which they exhibit. Sir Charles Sherrington in his Gifford Lectures has very vividly described the action of this vast assembly

[1] *The Study of Instinct*, Oxford University Press, 1951.

of resonating electrical circuits.[1] He speaks of it as an 'enchanted loom where millions of flashing shuttles (i.e. nerve impulses) weave a dissolving pattern, always a meaningful pattern, though never an abiding one; a shifting harmony of sub-patterns'. As for the details of behaviour of these many shuttles, we can use Leibniz' famous phrase, that 'everything which takes place within my mind is as mechanical as what goes on inside a watch', though perhaps the word 'mechanical' should be extended to include 'electrical'. Within this description what place can possibly be found for mind, as we are accustomed to regard it? The answer is that mind is a concept which we introduce, like other concepts such as gravitation, to make sense of our experiences. But these experiences are of many kinds, and if we use one single word 'mind' to relate them together, we must not be surprised at confusion resulting. There are physico-chemical questions about mind, which will have physico-chemical answers: and artistic questions to receive artistic answers: and spiritual questions, with appropriate answers also. All the different sets of answers are like different sections of our building: all may be true, none is exhaustive. We must use them in their proper context, and not waste valuable time and effort in the quite useless mixing of categories. . . .

Much the same can be said about my second 'famous debate', that between free-will and determinism.[2] As Max Planck pointed out several years ago,[3] this is as much a phantom problem as that between mind and matter. Let us think of Julius Caesar about to cross the Rubicon. The historian giving us this part of Roman history, and making use of the best scientific research into the situation of the time, will speak of political issues and innate temperament which effectively compelled Caesar's decision to burn his boats. Indeed, our historian is accounted a good historian just in so far as he can make the decision appear inevitable. But,

[1] *Man on his Nature*, Cambridge University Press, 1940.
[2] I have developed this at greater length in my Riddell Lectures, *Christianity in an Age of Science*, Oxford University Press, 1953, p. 22.
[3] *A Scientific Autobiography*, Williams & Norgate, 1950.

S.A.W.—M

with no less validity, we may try to project ourselves into Caesar's mind. One thing and one thing only, stands out in sharp relief: he has to decide. Political eventualities are the material of his decision, to be borne in mind and weighed up: but they do not themselves force him either way: at most they urge this or that action. It is a plain disregard of evidence and experience to deprive him of his moment of anxiety and decision. So what can we say of all this? Observed from without, the will is causally determined; observed from within it is free. The difference lies in the point of view (the section of the building) for no answer at all can be given until we have specified explicitly the viewpoint of our observation, and said whether we are actor or spectator. For man as actor the best concept is freewill: but for man as spectator it is determinacy. Once again let us be glad for the new development. As always the truth has made us free, but within a wider framework than could have been possible before. . . .

The last of my three phantom problems is that of teleology – whether, in Charles Kingsley's words we are obliged to choose between 'the absolute empire of accident and a living, immanent, ever-working God'. With all respects to the author of *Madame How and Lady Why* I believe that this choice is one that we make at our absolute discretion: for the two accounts of the development of the natural order are not to be regarded as exclusive. For some purposes one of these is better than the other, but in other circumstances the reverse will hold. . . . I think, for example, that if we wanted, we could describe evolution in those terms which I quoted earlier in this chapter: '*Homo sapiens* is simply a survival from the Neolithic Age';[1] but in doing so we must be careful not to use the adverb 'simply' to imply that no other description is valid. In actual fact, I don't believe that this particular description is very helpful, nor that it does justice to

[1] 'In the long run of evolution *Homo sapiens* is simply a survival from the Neolithic Age . . . civilization is nothing more than the accumulation of experience and knowledge, it reflects nothing other than the use to which man has put his brain.' *Man and His Gods*, by Homer Smith; Jonathan Cape, 1953.

the pattern even of technical biology. I am more impressed by the words of Sherrington[1] when he is discussing the evolution of cells in a human body and writes: 'It is as if an immanent principle inspired each cell with knowledge for the carrying out of a design' . . . Whether we use the language of teleology or not is a matter for our choice. But unless we do, we shall miss part of the pattern of nature shown by science. . . .

We have now reached the condition of recognising the validity of different conceptual patterns associated with substantially the same phenomena. . . . If this is true even in science itself, how much more true must it be when we are concerned with as wide a field as science and religion. And it is particularly true when we are discussing human motives and human thoughts. . . . The astronomer who turns from his telescope and exclaims: 'I swept the heavens and found no God'; and the man who, after focusing his microscope, rises with the exclamation, 'I have examined the brain and found no traces of love'[2] are both right, although both God and love can exist within the pictures that they see. We need to remember that Laplace, asserting that he 'had no need of that hypothesis' (God), and Descartes, crying 'Give me matter and motion, and I'll construct the universe', were both professing Christians; and, within certain limits, they spoke correctly. . . .

Each section of that physics laboratory in London is necessarily a two-dimensional affair. It is an abstraction of certain elements from the totality. If we call it a representation of the laboratory, then it is obviously partial. And however perfect we make our one single drawing it can never give us a satisfactory description of the building. Indeed, the greater technicalities of the diagram may sometimes serve to impede our sense of the total edifice, in much the same way that wearing blinkers helps the horse to see clearly what is directly in front at the expense of narrowing his field of view. But the building does become three-dimensional when we can place ourselves in the attitude to accept more than one 'view'. This, of course, is exactly what we do with our eyes

[1] *Man on His Nature.*
[2] J. W. Rowntree, *Claim your Inheritance*, Bannisdale Press, 1949.

in stereoscopic vision. We gain the sense of perspective and solidity and distance just because our two eyes see things slightly differently, and we accept the two accounts. We do not super-impose them – for that would make nonsense of what we see – but in some sense we do hold them together. I believe that some-thing akin to this is necessary if we would find God. In some stereoscopic way we must build up out of the imperfect abstrac-tions of any one conceptual framework, something of the full three-dimensional character of the reality which we continually encounter in every experience.

C. A. COULSON, *Science and Christian Belief*, 1955 (McNair Lectures, 1954)

And if we now grant Freud his contention that both our need for God and our conception of God have their seeds in our earliest relationships with our parents, does this in any way prove that God does not exist? And if we grant Coulson his claim that no single human viewpoint can suffice to comprehend the true nature of life and the universe, and that for this we need a superhuman viewpoint, is that any proof that a superhuman viewpoint exists?

Points for Discussion and Suggestions for Writing

Chapter One

1. Guided solely by text and footnote, draw a diagram of the system devised by Ptolemy to reduce a single planet's movements to a combination of circular ones. Compare with one another: (*a*) using the diagram as an aid in explaining the system to an audience, (*b*) writing the explanation which would need to accompany the diagram in a book on the subject, and (*c*) trying to explain the system without the aid of a diagram, as if to a radio audience.

2. What different aspects of science are represented by the con-contributions made respectively by Copernicus and Tycho Brahe to Newton's final solution of planetary motion?

3. Consider carefully the claim, and your reaction to it, that Shakespeare and Newton are the two most *creatively imaginative* Englishmen ever to have lived.

4. Bearing in mind Russell's remarks on absolute motion, define a 'receding galaxy'.

5. 'Some scientists found this picture profoundly unsatisfying', says Davy of the 'big bang' theory. 'It raises unanswerable questions about the origin of the universe.' Is the fact that a theory raises such questions a good reason for discarding it? Are there reasons why others might find the theory of 'continuous creation' equally 'unsatisfying' or unsatisfactory? Which do you

find the more satisfying, and does the kind of reason you have for preferring one or the other have any place in science?

6. Note the rhythms to Russell's sentences. Compare them with those of Taylor's, Bronowski's, or Davy's.

> All motion is relative,
> and there is no difference
> between the statements:

| the earth rotates once a day | and | the heavens revolve about the earth once a day. |

> The two mean exactly the same thing,

> just as it means the same thing if I say that a certain length is

| six feet | or | two yards. |

*   *   *

> Astronomy is easier

| if we take the sun as fixed | | than if we take the earth (as fixed), |

> just as accounts are easier

| in a decimal coinage | | (than in. . . .) |

*   *   *

| All motion is relative, | and it is a mere convention to take one body as at rest. |
| All such conventions are equally legitimate, | though not all are equally convenient. |

7. Examine closely: (*a*) Davy's use of analogy, and (*b*) his progress step by step through an argument in his concluding nine paragraphs.

Chapter Two

1. Consider the spread of meanings covered by *element(s)*, *elemental*, and *elementary*.

2. Use language as simple and non-technical as Hooke's to describe all the everyday properties of a fluid, as if you were a child, or a visitor from another planet, or for some other reason were discovering such properties for the first time.

3. Assuming the relative weights in the mass of the hydrogen and oxygen making up water to remain at $1 : 8$, what would their relative atomic weights have been had water turned out to be a quaternary compound (H_3O)?

4. 'Originally atoms were introduced to explain matter qualitatively through their movements and structure.' What exactly does Heisenberg mean by this sentence? Why did atoms fail in this original task, and what have or has met with better success?

5. Does such success mean that (*a*) in theory, and (*b*) in practice, all sciences are now a branch of physics?

6. What differences are there between the vocabulary Hooke used and that employed by Dalton or more recent scientists? What has been gained and/or lost?

Chapter Three

1. What is it which each of the analogies – machine or factory, whirlpool, population – tells us about the living body? Does the whirlpool analogy invalidate the machine one, and the population analogy invalidate the whirlpool one?

2. What does Young mean by 'Biology, like physics, has ceased to be materialist'? Would Huxley have been likely, living in the nineteenth century, to have understood such a point of view?

3. Write a paragraph comparing the living body with one of the following so as to throw some light on its workings:
    a bank, a nation, a field.

Chapter Four

1.

One electron	= nothing.
Several electrons, etc.	= a table.
(Many electrons, etc.	= a world.)
One letter	= nothing.
Several letters	= a word.
Many letters	= a story.

Does this analogy of Eddington's, or do any of the other devices he uses, go any distance towards achieving what he is here arguing cannot be achieved?

2. Summarize as briefly as possible the facts contained in the extract by Sherrington. What has been lost? What, if anything, gained?

3. Why should it be easier to reconcile (*a*) free will, (*b*) absolute standards in art, and (*c*) absolute standards in morals, with dualism than it is to reconcile them with any of the alternatives to dualism?

4. Is the straightforward impossibility of believing in it a good and sufficient reason for rejecting idealism?

5. Explain briefly why, according to Planck, the problem as to what relationship there is between body and mind just does not exist.

6. Deduce what you think you can of the personalities of the authors from the styles in which they write.

Chapter Five

1. Explain, as if to an intelligent ten- or eleven-year-old, what a vacuum is and why nature only appears to abhor one.

2. How would you have proceeded with the experiment, had you been Boyle, from the point where extract IV breaks off to a point where you would have been justified in considering Boyle's law to have been established?

3. List the respects in which Faraday might serve as a model for all scientific investigators.

4. What can we learn, from the extracts by Darwin and Wallace, about the similarities and differences between the paths followed by these two men in their search for a mechanism for evolution?

5. 'And I can remember the very spot on the road, whilst in my carriage, when to my joy the solution occurred to me', writes Darwin; and Wallace: 'Then there flashed upon me, as it had done twenty years before upon Darwin . . .' What do these phrases tell us about scientists and scientific method?

6. '*While* thinking . . . I *then* saw . . . Famine, droughts, floods . . . / *Then* there flashed . . . *Then* I at once saw . . . *But* this would only tend . . . / *But* along with Malthus . . . *Hence* it became obvious . . . The succession of . . .' These are the beginnings of Wallace's sentences. Note how the words in italics lead the thought on from sentence to sentence, even paragraph to paragraph.

7. Compare Boyle and Faraday as writers-up of experiments. How aware is one of the passive voice in reading Faraday?

Chapter Six

1. Employ the usefulness of scale models in engineering as an analogy to explain the usefulness of analogies.

2. Use an analogy, first as a means of *explaining*, then as a means of *describing*, one of the following:

(*a*) the circulation of the blood;
(*b*) the social life of bees or ants;
(*c*) the rise and fall of nations;
(*d*) marriage.

3. Write briefly on the importance of likenesses in human thought.

Chapter Seven

1. Compare your attitudes to theological predestination and scientific determinism.

2. If we accept that human beings are not free to act as they choose, what difference should this make if any to the way our society is run?

3. Does Ptolemy have the last laugh after all (chapter I, extract II)?

4. Does Bronowski suggest a causal connexion between the freedom of electrons and the freedom of men? Or are they, like his molecules in a gas or accidents on the road, merely an analogy?

5. Do you feel this 'pessimism which comes from our divided loyalties: our reverence for machines and, at odds with it, our nostalgia for personality'? If so, does Bronowski lend you a 'new optimism'?

Chapter Eight

1. Use the theory of natural selection to show how the evolution of the human brain must have depended on the prior existence of the human hand, just as the evolution of the human hand must have depended on the prior existence of the human brain.

2. Let the change from totipotent to differentiated cells be an analogy to illustrate something or other.

3. Which is more 'undifferentiated', a juke-box or a gramophone?

4. Is Rostand's attempt to draw a social lesson from biology as suspect, do you think, as those examples quoted by Medawar?

5. Omitting as much as possible of the purely illustrative material, summarize drastically the substance or argument first of the passage by Rostand, then of the one by Medawar. What differences are there between the two authors' methods of developing an argument? Have these anything to do with the fact that Medawar was writing a lecture to be broadcast to the general public?

Chapter Nine

1. Why have insects been able, in many cases, to survive as species if not as individuals the ravages of insecticides intended for their destruction better than Rachel Carson fears man himself may be able to?

2. To what other kinds of dangers brought about by himself, besides those dealt with in this chapter, may man find it difficult to adjust?

3. Write an extract from a novel about the future in which man has been unable to adapt to just one such change brought about by himself in himself or his environment.

4. Has Medawar no answer for Russell?

Chapter Ten

1. Explain, in as simple terminology as possible and preferably in one sentence each, the nature of *mass, acceleration*, and the *Second Law of Thermodynamics*.

2. What is the difference between an intellectual culture and an anthropological one? How apt is it to describe the scientific culture as being anthropological?

3. What novels have there been, besides C. P. Snow's, which deal with science and scientists? Why so few?

4. Why should scientists tend to prefer music to the other arts?

5. Does the non-scientist really lose as much by having little or no knowledge of science as the scientist loses by having little or no experience of the arts and humanities?

6. Do we in this country specialize too early in our schools and universities? If so, how can this be remedied?

7. Is it really possible, or worth while, to bridge the gap between scientist and non-scientist?

8. What does Jacquetta Hawkes mean by: 'The thalamic brain is feminine in its nature, in contrast with the essential masculinity of the Cerebrum'? Do you approve of the use of terms like 'masculine' and 'feminine' in a context like this?

9. Would it be the Thalamus or the Cerebrum of the person in question which interested you more in choosing, or being blessed with, each of the following: a teacher or professor, a headmaster, a medical specialist, a family doctor, a wife or husband, a Prime Minister, a son or daughter?

10. Do you agree that a person can be highly intelligent and yet lacking in wisdom, even silly?

11. Is it true that a people is mainly remembered and judged by its works of art? Is this as likely to be, or seem, true in the future as in the past?

12. Be an historian of the twenty-second century and describe our present civilization, bringing out those features of our achievement for which you think we most deserve to be remembered.

Chapter Eleven

1. Myth, as everyone knows, is a method of explaining, or rather accounting for ('explaining' implies too conscious and calculating a process) puzzling natural events in supernatural or anthropomorphic terms. It may also, like legend but at a further remove, be a way of preserving something in a tribe's or society's remote past; and it can be used to enlist divine approval for, or to enshrine, particular policies or ways of behaviour. Yet again, it may be a means of expressing in simple and concrete terms those vast abstract truths about the universe which are to be apprehended otherwise only with great difficulty and in the shadowiest kind of way. Often, indeed most often, it is a mixture of these, or one thing to one man and another to another. Which is it in the following instances?

(*a*) Thor and Vulcan, blacksmiths by appointment to Odin and Jupiter respectively.

(*b*) A primitive matriarchal society (probably one ignorant of the male role in procreation) is overrun by a more advanced patriarchal one. Their resultant and compositive mythology makes 'the creation of the world begin precisely with a male god triumphing over a female goddess'.

(*c*) The Polynesians, some of whose ancestors if one is to believe

Thor Heyerdhal came from South America, trace their descent from Tiki, son of the sun, who came from the East.

(*d*) God forbade Adam and Eve to eat of the fruit of the tree of knowledge; when they disobeyed him they were cast out of Eden.

2. Is a truth less of a truth because enshrined in an outdated mythology? Is an outdated mythology less outdated because it enshrines a truth? (Think of actual instances.) Can one simultaneously believe in a dogma (myth?) and be aware that it may need to be brought up to date at any time?

3. Why, according to Freud, do Christians think of God as a Father? Why, according to Christians themselves, do they do so? Are the two explanations mutually exclusive?

4. What, from this passage, would you imagine Freud's relationship with his own father to have been like? (Check the accuracy of your estimate after having made it.) Have considerations like this any valid bearing on Freud's argument?

5. 'This does not mean that light is *both* corpuscle and wave: the dualism lies not in what Kant would call "the-thing-in-itself", but in our interpretation of it. . . .' Explain what is meant to someone who is unable, or unwilling, to follow Coulson at this point in his argument.

6. Expound at greater length Tinbergen's three ways of studying instinct. Of the three phantom problems which Coulson has yet to discuss (mind/body, free-will/determinism, and teleology), how many are already rearing their ugly heads?

7. In which previous passages have these phantom problems arisen? Is Coulson in agreement in all cases with the authors in question? Compare his handling of the problems (somewhat abbreviated here) with theirs.

8. List and examine the situations to which Coulson applies his 'underground laboratory' analogy. Are the uses he makes of it equally happy or appropriate in all cases?

9. 'In some stereoscopic way we must build up out of the imperfect abstractions of any one conceptual framework, something of the full three-dimensional character of the reality which we continually encounter in every experience.' COULSON, p. 180.

'But reminding us that the step from electrical disturbance in the brain to the mental experience is the mystery it is, the mind adds the third dimension when interpreting the two-dimensional picture! Also it adds colour; in short it makes a three-dimensional visual scene out of an electrical disturbance.' SHERRINGTON, p. 62.

'It is all symbolic, and as a symbol the physicist leaves it. Then comes the alchemist Mind who transmutes the symbols. The sparsely spread nuclei of electrical force become a tangible solid, their restless agitation becomes the warmth of summer; the octave of ætherial vibrations becomes a gorgeous rainbow. Nor does the alchemy stop here. In the transmuted world new significances arise which are scarcely to be traced in the world of symbols; so that it becomes a world of beauty and purpose—and, alas, suffering and evil.' EDDINGTON, p. 58.

How feasible, and how helpful, would it be to make much more than Coulson does of his analogy between what the mind does with nerve impulses from our senses, and what religious belief can do with our ideas about the universe around us?

Suggestions for Further Reading

Histories of Science and General Background

BUTTERFIELD, H. *The Origins of Modern Science*, 1300–1800. Bell 1949 (New Edition 1957). Scientific development as seen by a historian in relation to what else was happening at the time.

GRANADA TV NETWORK. *Discovery: Developments in Science.* Methuen 1961. Popular but far from superficial studies of various modern developments in science.

PLEDGE, H. T. *Science Since 1500.* H.M.S.O. 1939 (Harper Torchbooks 1959). Scholarly and detailed – gnomic style.

SINGER, CHARLES. *A Short History of Scientific Ideas to 1900.* O.U.P. 1959. Comprehensive and readable.

Chapter One

HOYLE, FRED. *Frontiers of Astronomy.* Heinemann 1955 (Mercury 1961).

KOESTLER, ARTHUR. *The Sleepwalkers.* Hutchinson 1959. Contentious and interesting history of astronomy up to Galileo.

LOVELL, Sir BERNARD. *The Individual and the Universe* (Reith Lectures 1958). O.U.P. 1959.

PAYNE-GAPOSCHKIN, CECILIA. *Introduction to Astronomy.* Eyre & Spottiswoode 1956 (University Paperbacks 1961). Historical survey, followed by scholarly summary of present-day views.

TOULMIN, STEPHEN, and GOODFIELD, JUNE. *Fabric of the Heavens.* Hutchinson 1961. History of astronomical ideas.

WHITROW, G. V. *The Structure and Evolution of the Universe.* Hutchinson 1959. Needs some knowledge of mathematics for full understanding.

Chapter Two

FINDLAY, ALEXANDER. *General and Inorganic Chemistry.* Methuen 1953. Good book for those needing to revise what happened to chemistry after Dalton.

FRIEND, J. NEWTON. *Man and the Chemical Elements.* Griffin 1952. Informal history of chemistry.

HECHT, SELIG. *Explaining the Atom* (rev. edn. with additional chapters by Eugene Rabinowitch). Gollancz 1955. Lucid account of changing views on the nature of matter.

JONES, G. O., ROTBLAT, J., and WHITROW, G. J. *Atoms and the Universe.* Eyre & Spottiswoode 1956 (2nd rev. edn. 1962). Atoms in relation to classical physics, nuclear physics and astrophysics.

KENDALL, JAMES. *At Home Among the Atoms.* Bell 1929. A not too technical account of the relationship between atoms and chemistry.

OPPENHEIMER, J. R. *Science and the Common Understanding* (Reith Lectures 1953). O.U.P. 1954.

THOMSON, Sir GEORGE. *The Atom.* O.U.P. 5th edn. 1956.

WILSON, WILLIAM. *The Microphysical World.* Methuen 1951. Useful summary – not too advanced.

Chapter Three

ABRAHAMS, Sir ADOLPHE. *The Human Machine.* Penguin 1956.

ASIMOV, ISAAC. *The Chemicals of Life.* Bell 1956. Simple biochemistry.

BALDWIN, ERNEST. *The Nature of Biochemistry.* C.U.P. 1962.

BECK, WILLIAM S. *Modern Science and the Nature of Life.* Penguin 1961 (N.Y., Harcourt Brace, 1957). Excellent account of the latest views on the nature of life, coupled with the author's views on the nature of science.

BONNER, J. T. *Cells and Societies.* O.U.P. for Princeton U.P. 1955.

BOREK, ERNEST. *Man the Chemical Machine.* O.U.P. for Columbia U.P. 1952.

BUTLER, J. A. V. *Inside the Living Cell.* Allen & Unwin 1959.

READ, JOHN. *A Direct Entry to Organic Chemistry.* Methuen 1953. Admirably lucid book on the more elementary aspects of chemistry as a background to life.

Chapter Four

BUTLER, J. A. V. *Man is a Microcosm.* Macmillan 1950. Biochemistry, concluding with a discussion of mental activity.

LASLETT, PETER (Ed.). *The Physical Basis of Mind.* Blackwell 1950. A symposium.

RYLE, GILBERT. *The Concept of Mind.* Hutchinson 1949 (Penguin 1964). A modern philosopher looks at the question.

SCHRÖDINGER, E. *What is Life?* C.U.P. 1944. Can physics and chemistry account for it all? Yes, says Schrödinger.

Chapter Five

ARBOR, AGNES. *The Mind and the Eye.* C.U.P. 1954. A biologist writes on scientific method.

BECK, STANLEY D. *The Simplicity of Science.* Penguin 1962 (N.Y. Doubleday 1959).

BEVERIDGE, W. I. B. *The Art of Scientific Investigation.* Heinemann 1953.

CONANT, JAMES B. *Science and Common Sense*. O.U.P. for Yale U.P. 1951. Strategy and tactics of some great investigators.

KENDALL, JAMES. *Michael Faraday, Man of Simplicity*. Faber 1955.

TATON, R. *Reason and Chance in Scientific Discovery*. Hutchinson 1957.

TOULMIN, STEPHEN. *The Philosophy of Science*. Hutchinson 1953.

WIGHTMAN, W. P. D. *The Growth of Scientific Ideas*. Oliver & Boyde 1950. The story of the development of selected dominant ideas in science.

Chapter Six

ABRAHAMS, Sir ADOLPHE. *The Human Machine*. Penguin 1956.

BONNER, J. T. *Cells and Societies*. O.U.P. for Princeton U.P. 1955.

HUTTEN, E. H. *The Language of Modern Physics*. Allen & Unwin 1956.

SAVORY, T. H. *The Language of Science*. Deutsch 1953.

Chapter Seven

EDDINGTON, Sir ARTHUR. *New Pathways in Science*. C.U.P. 1934.

EINSTEIN, ALBERT. *Relativity (A Popular Exposition)*. Methuen 1920 (15th enlarged edn. 1954, University Paperbacks 1960).

EINSTEIN, ALBERT, and INFELD, LEOPOLD. *The Evolution of Physics*. C.U.P. 1938.

HANSON, NORWOOD RUSSELL. *Patterns of Discovery*. C.U.P. 1958.

HEISENBERG, WERNER. *Philosophical Problems of Nuclear Science*. Faber 1952.

JEANS, Sir JAMES. *The New Background of Science*. C.U.P. 2nd ed. 1947.

LEVINSON, H. C. *The Science of Chance*. Faber 1952. The mathematics of chance.

PEIETRIS, R. E. *The Laws of Nature*. Allen & Unwin 1955.

Chapter Eight

BONNER, J. T. *The Ideas of Biology*. Eyre & Spottiswoode 1963.

EISELEY, LOREN. *Darwin's Century*. Gollancz 1959. A history of evolutionary ideas by a biologist.

FORD, E. B. *Mendelism and Evolution*. Methuen 1957. A good summary of present day views on evolution.

GREENE, JOHN C. *The Death of Adam*. Mentor 1961 (Iowa State U.P. 1959). A very fully documented history of the growth of evolutionary ideas by a historian.

HUXLEY, JULIAN. *Evolution in Action*. Chatto & Windus 1953. Evolution in the past and the future.

HUXLEY, T. H., and HUXLEY, JULIAN. *Evolution and Ethics* (Romanes Lectures of 1893 & 1943). Pilot Press 1948.

LACK, DAVID. *Evolutionary Thought and Christian Belief*. Methuen 1957. Very fair to both sides.

SIMPSON, GEORGE G. *The Meaning of Evolution* (Terry Lectures). O.U.P. 1950, rev. & abridged edn. Muller (Mentor Books) 1956. Evolution in relation to human values and purposes.

TEILHARD DE CHARDIN, PIERRE. *The Phenomenon of Man*. Collins 1959. A religious, almost mystical, view of evolution.

Chapter Nine

ANDERSON, M. S. *Geography of Living Things*. E.U.P. 1951. Man and his geographical environment – very readable.

HARRISON, GEORGE RUSSELL. *What Man May Be*. Cassell 1957.

JACKS, G. V. *Soil*. Nelson 1954.

RUSSELL, BERTRAND (Earl RUSSELL). *The Impact of Science on Society*. Allen & Unwin 1952.

RUSSELL, Sir E. JOHN. *Science and Modern Life* (Beckly Lecture 1955). Epworth Press 1955.

Chapter Ten

BRONOWSKI, J. *Science and Human Values*. Hutchinson 1961.

CALDER, RITCHIE. *Science Makes Sense*. Allen & Unwin 1955.

EASTWOOD, W. (Ed.). *Science and Literature*. Macmillan 1957. An anthology of science in literature.

LEAVIS, F. R. and YUDKIN, M. *Two Cultures? The Significance of C. P. Snow* (Richmond Lecture 1962). Chatto & Windus 1962. Leavis' vitriolic rejoinder to Snow.

Chapter Eleven

DAMPIER, Sir WILLIAM. *A History of Science and its Relations with Philosophy and Religion*. C.U.P. 4th edn. 1948.

KLINE, MORRIS. *Mathematics in Western Culture*. Allen & Unwin 1954.

LACK, DAVID. *Evolutionary Thought and Christian Belief*. Methuen 1957.

RAVEN, CHARLES E. *Natural Religion and Christian Theology* (Gifford Lectures, 1951 & 1952). C.U.P. 1953 (2 vols.). 1st Series, Science and Religion, 2nd Series, Experience and Interpretation.

WEIZÄCKER, C. F. *The History of Nature*. Routledge 1951. Mainly on the development and evolution of the universe, and on life from a scientific point of view, but asks whether science can encompass ethics.

Biographical Notes

ANDRADE, E. N. da C., F.R.S., born 1887. Professor of Physics at London University 1928–50; Director of the Royal Institution and the Davy-Faraday Laboratory 1950–52; well-known for work on atomic structure.

ARISTOTLE, 384–322 B.C. Greek Philosopher and Naturalist.

BACON, Sir FRANCIS, 1561–1626. Philosopher and Statesman; champion, in *Advancement of Learning* (1605) and *Novum Organum* (1620), of the inductive as opposed to the deductive method of reasoning, and of the need to accumulate fact and observation.

BOYLE, The Hon. ROBERT, 1627–1691. One of the founder Fellows of the Royal Society, best known for his law relating volume and pressure of a gas, and for his definition of an element as the practical limit of chemical analysis.

BRONOWSKI, J., born 1908. Director-General of the Process Development Department of the N.C.B. at Cheltenham, and one-time University Lecturer; Mathematician and Scientist who has written books on the poetry of Blake, Shelley and others, and several radio plays, including *The Face of Violence* which won the Italia Prize in 1959.

CARSON, RACHEL, 1907–1964. Writer and Naturalist; served with the U.S. Fish and Wildlife Service 1936–52; author of *The Sea Around Us* and other scientific best sellers.

CLARK-KENNEDY, A. E. Physician to the London Hospital 1928–58, and Dean of the London Hospital Medical College 1937–53.

CONANT, JAMES B., born 1893. Professor of Organic Chemistry at Harvard 1928, and President of the University 1933–53. U.S. High Commissioner to W. Germany 1953–55, and Ambassador 1955–57.

COULSON, C. A., F.R.S., born 1910. Professor of Mathematics at Oxford since 1952; formerly Professor of Theoretical Physics, Kings College, London, and I.C.I. Fellow in Chemistry at Oxford. Vice-President of the Methodist Conference 1959.

DALTON, JOHN, 1766–1844. Worked as a farm labourer; then ran a school with his brother. Appointed Teacher of Mathematics and Physical Science at New College, Manchester, in 1794; the first to relate atoms to chemical processes, and to investigate the effect of temperature on vapour pressure.

DARWIN, CHARLES ROBERT, 1809–1882. Served as Naturalist aboard *H.M.S. Beagle* during a voyage round South America in 1831–32; the observations he made on that voyage led eventually to *The Origin of Species by Means of Natural Selection* in 1859.

DAVY, JOHN, born 1927. Science correspondent of the *Observer*, and broadcaster.

DE BROGLIE, 7th Duc LOUIS, born 1892. Professor of Physics at Paris 1932; the first to link wave and corpuscular theories in nuclear physics; Nobel Prize for Physics 1929.

DEMOCRITUS, *c.* 460–370 B.C. Greek Philosopher who, with his older contemporary Leucippus, postulated the first atomic theory; of the many books he wrote only a few fragments survive.

EDDINGTON, Sir ARTHUR, F.R.S., 1882–1944. Professor of Astronomy and Director of the Observatory at Cambridge; a great popularizer of new scientific concepts.

FARADAY, MICHAEL, 1791–1867. Worked for a bookbinder, read scientific books, attended Sir Humphry Davy's lectures, and became his assistant in 1813. Appointed Professor of Chemistry at the Royal Institution 1833; chiefly famous for his work on electricity and magnetism.

FREUD, SIGMUND, 1856–1939. Professor of Neuropathology in Vienna 1902; a refugee from the Nazis living the last year of his life in London. Originator of psycho-analysis, whether by his own method or by those of his divergent disciples, Jung and Adler.

HARRISON, GEORGE RUSSELL, born 1898. Professor of Physics 1930, and Dean of Science 1942, at the Massachusetts Institute of Technology.

HAWKES, JACQUETTA, born 1910. Archaeologist and Author, best known for *A Land*, 1950.

HEISENBERG, WERNER, F.R.S., born 1910. Professor of Physics and Director of the Max Planck Institute for Physics and Astro-Physics at Munich since 1958; has held similar posts in Leipzig, Berlin and Göttingen. Most famous for work on quantum theory and on atomic structure; enunciated the famous uncertainty principle which bears his name.

HOOKE, ROBERT, 1635–1703. Worked as Boyle's assistant for a time; an early user of the microscope, he discovered the existence of vegetable cells; he also discovered Hooke's law – namely that the extension of a spring is proportional to the force applied to it – and experimented widely in many other directions.

HUXLEY, Sir JULIAN, F.R.S., born 1887. Biologist and Writer; Professor of Zoology at Kings College, London, 1925–27, and at the Royal Institution 1926–29; first Director-General of U.N.E.S.C.O. 1946–48. Grandson of T. H. Huxley.

HUXLEY, T. H., 1825–95. Famous principally as an early convert to Darwin's theory of natural selection, and thereafter its most trenchant advocate and effective popularizer.

JACKS, G. V., born 1901. Director of Commonwealth Bureau of Soils since 1946; Editor of *Journal of Soil Science* 1949–61.

JEANS, Sir JAMES, 1877–1946. Mathematician and Physicist; Professor of Astronomy at the Royal Institution and co-popularizer with Eddington of many new scientific concepts.

MEDAWAR, P. B., F.R.S., born 1915. Professor of Zoology at Birmingham 1947–51, and at University College, London, 1951–62; Nobel Prize for Medicine 1960.

NEWTON, Sir ISAAC, 1642–1727. Professor of Mathematics at Cambridge 1669; F.R.S. 1671; *Philosophiae Naturalis Principia Mathematica* 1687; M.P. 1689; Master of the Mint 1699; President of the Royal Society 1703; *Opticks* 1704. Having by the age of 45 made what has remained perhaps the greatest single contribution to science by a single human mind, he spent much of his later life in intensive research as to the precise significance of the prophetic chapters of the Book of Daniel.

PLANCK, MAX, F.R.S., 1858–1947. Professor of Physics at Berlin, 1889; Quantum Theory 1900; Nobel Prize for Physics 1918; father-figure, together with Einstein, of modern physics.

ROSTAND, JEAN, born 1894. Biologist and Writer; Member of the Académie Français.

RUSSELL, BERTRAND (3rd Earl), O.M., F.R.S., born 1872. Philosopher and Mathematician who, in conjunction with Whitehead, wrote *Principia Mathematica* (1910–13), an attempt to link if not unify Mathematics and Logic; Nobel Prize for Literature 1950.

SHERRINGTON, Sir CHARLES, O.M., F.R.S., 1857–1952. Professor of Physiology at Liverpool and Oxford; leading authority on the nervous system; President of the Royal Society 1920–26; Nobel Prize for Medicine 1932.

SNOW, Sir CHARLES P., born 1905. Began his career before the war as Fellow and Tutor of Christ's College, Cambridge; moved to the Civil Service, where he helped to direct the nation's scientific war-effort; has written eleven novels to date in the *Strangers and Brothers* sequence, and an early novel *The Search* which is largely concerned with the nature, the fascination, and the temptations, of scientific research.

TAYLOR, F. SHERWOOD, 1897–1950. Chemist and Writer; Director of the Science Museum, South Kensington; one-time School Master and University Lecturer.

WALLACE, ALFRED RUSSEL, 1823–1913. Naturalist who spent many years studying life in Malayan jungles and elsewhere; arrived independently and with fewer hesitations at the same conclusion – viz. that life had evolved to its present forms by means of natural selection – as Darwin had reached some years earlier. They presented simultaneous papers on the subject in 1858. In later years Wallace held more firmly than Darwin to the exclusive importance of natural selection, rejecting all compromise with Lamarckianism.

WALTER, W. GREY, born 1910. Director of the Physiological Dpt., Burden Neurological Institute, Bristol, since 1939. Pioneer worker on Electroencephalography and Neurophysiology.

WHITEHEAD, A. N., F.R.S., 1861–1947. Mathematician and Philosopher, holding academic posts as the former in Cambridge and London, and the latter in Harvard. Collaborated with Bertrand Russell in *Principia Mathematica*.

YOUNG, J. Z., F.R.S., born 1907. Professor of Anatomy in London University since 1945.

Index